DRY FIRE

Handgun Skills

Logbook

Shawn Kovacich

AUTHOR APPROVED SHOOT/NO SHOOT EDITION

Printed in the United States of America

Library of Congress Control Number: 2019912092

ISBN #978-1-7336809-4-3

Table of Contents

Disclaimer

Please note that the author, Shawn Kovacich, and the publisher, **Shoot-NO-Shoot Productions**, Chikara Kan, Inc., of this logbook are **Not Responsible** in any manner whatsoever, for any injury, which may occur by reading and/or following the instructions located within this logbook. The distributor and any and all other individuals and or businesses connected with the production and/or distribution of this logbook also disclaim any and all liability for any damage or injury of any kind that may result from the appropriate use or inappropriate misuse of any information contained within this logbook.

The techniques that you will be executing while utilizing this logbook are potentially lethal in nature and are intended to be used in a **safe and law abiding manner** during practice and also if/when confronted with a potentially life threatening situation that may or may not require the use of lethal force in order to protect ones life. It is the reader or readers direct responsibility to research and comply with all applicable laws regarding the use of force in a self-defense situation in the areas in which they reside, do business, or frequent. **Ignorance of the law never has been, nor is it likely that it ever will be, a viable defense.**

Due to the fact that the techniques described in this book are physical in nature, it is essential that the reader or readers of this book consult a qualified and competent physician before following any of the activities, physical or otherwise, that you will be executing while utilizing this logbook.

This book and all the information contained therein is intended to be used as a supplemental training aid, and should be used as such, under the guidance of a qualified and competent firearms instructor.

This book and the information contained within are for informational purposes only!

Copyrights & Trademarks

SAFETY

Due to the physical nature of the activities that you will be participating in while utilizing this logbook, it is highly recommended that you first consult a qualified and competent physician before starting any physical training routine. Regardless of your current physical condition. Remember, an ounce of prevention is worth a pound of cure. And also remember, **Safety First, Foremost, and Always!**

Additionally, make sure that you have shooting glasses for your eyes and sound deadening ear muffs or plugs for your ears! Make sure that you know, understand, and abide by **ALL SAFETY RULES** at whatever range you are practicing at, regardless of your experience. Remember, you can never be too safe.

AND FINALLY...

You would do well to remember that progress should be measured in inches, not in miles. You are not going to execute 10 or 20 dry-fire drills and then all of a sudden be showing a remarkable improvement in your shooting skills. It takes time, proper practice, full immersion and active participation, and consistency in practice in order to achieve positive results. You probably won't see any big changes after just a few dozen repetitions of a particular dry-fire drill, but you should be seeing some improvement after the first 100 repetitions of a particular dry-fire drill. And, you should continue to show improvement after every 500 repetitions of a particular dry-fire drill that you execute, if you have fully immersed yourself both physically and mentally, while actively participating in each and every training session.

In addition to that, if you have also been studying and refining your dry-fire and live-fire handgun skills under the watchful eye of a qualified and competent firearms instructor, then you will make even greater progress in honing your shooting skills.

HOW TO USE THIS LOGBOOK

In order to obtain the full benefits of this **Dry Fire: Handgun Drills Logbook**, you will want to follow these simple instructions on how to best utilize this logbook. This logbook should be used along with your active participation with the various dry-fire drills that you are currently practicing. These dry-fire drills that you have learned and/or are learning, should be practiced on a regular and consistent basis while keeping an extensive and detailed record of the number of repetitions that your are executing during each practice session.

However, please make sure that you are first doing the dry-fire drills **SAFELY** and correctly before you start keeping track of the hundreds, and perhaps even thousands, of repetitions that you are going to start keeping track of. Because if you are doing these dry-fire drills incorrectly the first time, and you continue to practice them incorrectly, how do you think you are going to execute them on the 1,000th time?

COVER & SPINE

When you look at the front cover and the spine, you will see a small white box in the lower left hand corner on the front cover and a smaller white box on the spine. These are include in order to give you a space in which to write down the volume number or other such indicator of the logbook that you are currently using.

HOLSTER POSITION

How are you carrying your firearm? Are you carrying on your strong side hip? And if so, are you carrying it in a OWB (outside the waistband) holster? Or perhaps you are carrying on your strong side hip but in a IWB (inside the waistband) holster? You may also be carrying in a appendix carry holster, shoulder holster, cross draw holster, small of the back holster, a purse, fanny pack, a pocket holster, etc.

HAND

This is where you would enter the hand that you are using during this particular **"Dry-Fire"** drill. Are you using your dominate (**"Right"** or **"Left"**) or non-dominate (**"Left"** or **"Right"**) hand?

START DATE

This is where you will enter the **"Date"** that you started keeping track of the number of repetitions that you executed of this particular **"Dry-Fire"** drill. Be sure and include the month, day, and year.

SPARE MAGAZINE POSITION

This would be the location in which you are carrying your spare magazines.

FIREARM

This would be the make and model of the particular **"Firearm"** that you used during your dry-fire shooting drills. So if you used a **Glock 19**, then the make of the **"Firearm"** would be Glock and the model would be 19. If you used let's say a **Kimber 1911 Royal II**, then "Kimber" would be the make, while the model would be the "1911 Royal II."

COVER GARMENT

How are you concealing your firearm? What type of **"Cover Garment"** are you using to cover it? Jacket, shirt, etc. How is that **"Cover Garment"** fastened? Or, are you wearing it unfastened? Have you practiced drawing from various types of cover garments? What have you discovered? Does the same method of drawing work with all cover garments? Or, perhaps you are carrying in an off-body method such as a purse, briefcase, fanny pack, etc.

CALIBER

This would be the **"Caliber"** of bullet that is used in the firearm that you used during the live fire shooting drills. So, if you used a Glock 19, the caliber would be 9mm. If you used the Kimber 1911 Royal II, the caliber would be .45 ACP or 45 Auto.

NOTES

This section allows you to write down any particular **"Notes"** that you may have concerning this particular dry-fire drill that you are keeping track of.

SLOW REPS

This section allows you to keep track of the number of slow repetitions or **"Slow Reps"** that you executed throughout the time frame that you were also executing your regular repetitions or **"Reg Reps"** of this particular dry-fire drill. My personal recommendation is that you should execute at least 5 to 10 **"Slow Reps"** at the beginning and at the end of every session that you are working on this particular dry-fire drill.

For example, if you are going to execute 50 repetitions of a particular dry-fire drill at regular speed **"Reg Reps,"** then you should start out be executing 5 to 10 repetitions of that dry-fire drill at a much slower speed **"Slow Reps"** in order to check for correct technique, and efficient and effective movement. Follow this with up to 50 repetitions of the dry-fire drill at regular speed and then finish up with another 5 to 10 slow repetitions. This would give you a total of 60 to 70 repetitions of this particular dry-fire drill executed in this particular training session, 10 to 20 **"Slow Reps"** and 50 **"Reg Reps."**

REG REPS

The regular repetitions or **"Reg Reps"** section allows you to circle or check off the number of repetitions of this particular dry-fire drill that you are keeping track of in increments of 10. So, for every 10 repetitions that you execute of a particular dry-fire drill, you would check off the next 10 repetition increment. For example; if you are just starting out and you just finished executing say 50 repetitions of a particular dry-fire drill, you would then circle or check off the following numbers; 10, 20, 30, 40, and 50.

NOTES

This **"Notes"** section, allows you the opportunity to write down any particular notes that you may have concerning this particular dry-fire drill after every 1,000 repetitions executed. There is no notes section after the **"Slow Reps,"** as your slow repetitions are intended to be executed at various stages during the execution of the 5,000 regular repetitions or **"Reg Reps."** Be sure and include the month, day, and year. Use the **"Additional Notes"** section included at the end of this logbook to keep track of any additional items that you feel you need to keep track of during the course of utilizing this logbook.

END DATE

This is where you will enter the **"End Date"** in which you completed the 1,000 slow repetitions **"Slow Reps"** and the 5,000 regular repetitions **"Reg Reps"** of this particular dry-fire drill.

CODE KEY

This section was included in order to allow you the opportunity to personalize the way you keep track of your progress by allowing you to use a variety of symbols, ink colors, etc., in order to keep track of any particularities that you wanted to include in your own personal way to add to your training regiment. I have included some various possibilities that you could use, or come up with your own that may fit your own particular needs. So without further ado, let's get started.

BLACK INK = Executed this dry-fire drill under the maximum allotted time.

BLUE INK = Executed this dry-fire drill at the maximum allotted time.

RED INK = Executed this dry-fire drill over the maximum allotted time.

X (over the number) = Time, technique, mind-set, was optimal during this dry-fire drill.

CIRCLE (around the number) = Time, technique, mind-set, was average during this dry-fire drill.

SQUARE (around the number) = Time, technique, mind-set, was below average during this dry-fire drill.

TRIANGLE (around the number) = Refer to **"Additional Notes"** section for more details.

CODE KEY

Here is the **"Code Key"** for you to personalize to your own particular needs and training regiment. I would strongly recommend that you read through the **"How To Use This Logbook"** section prior to using this logbook in order to maximize its use to its fullest potential. Be sure and fill this out prior to starting your logbook.

BLACK INK ______________________________

____

BLUE INK ______________________________

____

RED INK ______________________________

____

X ______________________________

____

CIRCLE ______________________________

____

SQUARE ______________________________

____

TRIANGLE ______________________________

____

FIELD STRIPPING

Holster Position:______________ **Hand:** ______ **Start Date:** _________
Spare Magazine Position: ____________ **Firearm:** ____________________
Cover Garment: __________________________________ **Caliber:** ______
Notes:____

____

____

Slow Reps: 10 - 20 - 30 - 40 - 50 - 60 - 70 - 80 - 90 - 100 - 110 - 120 - 130
140 - 150 - 160 - 170 - 180 - 190 - 200 - 210 - 220 - 230 - 240 - 250 - 260
270 - 280 - 290 - 300 - 310 - 320 - 330 - 340 - 350 - 360 - 370 - 380 - 390
400 - 410 - 420 - 430 - 440 - 450 - 460 - 470 - 480 - 490 - 500 - 510 - 520
530 - 540 - 550 - 560 - 570 - 580 - 590 - 600 - 610 - 620 - 630 - 640 - 650
660 - 670 - 680 - 690 - 700 - 710 - 720 - 730 - 740 - 750 - 760 - 770 - 780
790 - 800 - 810 - 820 - 830 - 840 - 850 - 860 - 870 - 880 - 890 - 900 - 910
920 - 930 - 940 - 950 - 960 - 970 - 980 - 990 - 1000

Reg Reps: 10 - 20 - 30 - 40 - 50 - 60 - 70 - 80 - 90 - 100 - 110 - 120 - 130
140 - 150 - 160 - 170 - 180 - 190 - 200 - 210 - 220 - 230 - 240 - 250 - 260
270 - 280 - 290 - 300 - 310 - 320 - 330 - 340 - 350 - 360 - 370 - 380 - 390
400 - 410 - 420 - 430 - 440 - 450 - 460 - 470 - 480 - 490 - 500 - 510 - 520
530 - 540 - 550 - 560 - 570 - 580 - 590 - 600 - 610 - 620 - 630 - 640 - 650
660 - 670 - 680 - 690 - 700 - 710 - 720 - 730 - 740 - 750 - 760 - 770 - 780
790 - 800 - 810 - 820 - 830 - 840 - 850 - 860 - 870 - 880 - 890 - 900 - 910
920 - 930 - 940 - 950 - 960 - 970 - 980 - 990 - 1000
Notes: ____

____

____

Reg Reps: 10 - 20 - 30 - 40 - 50 - 60 - 70 - 80 - 90 - 100 - 110 - 120 - 130
140 - 150 - 160 - 170 - 180 - 190 - 200 - 210 - 220 - 230 - 240 - 250 - 260
270 - 280 - 290 - 300 - 310 - 320 - 330 - 340 - 350 - 360 - 370 - 380 - 390
400 - 410 - 420 - 430 - 440 - 450 - 460 - 470 - 480 - 490 - 500 - 510 - 520
530 - 540 - 550 - 560 - 570 - 580 - 590 - 600 - 610 - 620 - 630 - 640 - 650
660 - 670 - 680 - 690 - 700 - 710 - 720 - 730 - 740 - 750 - 760 - 770 - 780
790 - 800 - 810 - 820 - 830 - 840 - 850 - 860 - 870 - 880 - 890 - 900 - 910
920 - 930 - 940 - 950 - 960 - 970 - 980 - 990 - 1000
Notes: ____

____

____

Reg Reps: 10 - 20 - 30 - 40 - 50 - 60 - 70 - 80 - 90 - 100 - 110 - 120 - 130
140 - 150 - 160 - 170 - 180 - 190 - 200 - 210 - 220 - 230 - 240 - 250 - 260
270 - 280 - 290 - 300 - 310 - 320 - 330 - 340 - 350 - 360 - 370 - 380 - 390
400 - 410 - 420 - 430 - 440 - 450 - 460 - 470 - 480 - 490 - 500 - 510 - 520
530 - 540 - 550 - 560 - 570 - 580 - 590 - 600 - 610 - 620 - 630 - 640 - 650
660 - 670 - 680 - 690 - 700 - 710 - 720 - 730 - 740 - 750 - 760 - 770 - 780
790 - 800 - 810 - 820 - 830 - 840 - 850 - 860 - 870 - 880 - 890 - 900 - 910
920 - 930 - 940 - 950 - 960 - 970 - 980 - 990 - 1000

Notes: ______________________________

Reg Reps: 10 - 20 - 30 - 40 - 50 - 60 - 70 - 80 - 90 - 100 - 110 - 120 - 130
140 - 150 - 160 - 170 - 180 - 190 - 200 - 210 - 220 - 230 - 240 - 250 - 260
270 - 280 - 290 - 300 - 310 - 320 - 330 - 340 - 350 - 360 - 370 - 380 - 390
400 - 410 - 420 - 430 - 440 - 450 - 460 - 470 - 480 - 490 - 500 - 510 - 520
530 - 540 - 550 - 560 - 570 - 580 - 590 - 600 - 610 - 620 - 630 - 640 - 650
660 - 670 - 680 - 690 - 700 - 710 - 720 - 730 - 740 - 750 - 760 - 770 - 780
790 - 800 - 810 - 820 - 830 - 840 - 850 - 860 - 870 - 880 - 890 - 900 - 910
920 - 930 - 940 - 950 - 960 - 970 - 980 - 990 - 1000

Notes: ______________________________

Reg Reps: 10 - 20 - 30 - 40 - 50 - 60 - 70 - 80 - 90 - 100 - 110 - 120 - 130
140 - 150 - 160 - 170 - 180 - 190 - 200 - 210 - 220 - 230 - 240 - 250 - 260
270 - 280 - 290 - 300 - 310 - 320 - 330 - 340 - 350 - 360 - 370 - 380 - 390
400 - 410 - 420 - 430 - 440 - 450 - 460 - 470 - 480 - 490 - 500 - 510 - 520
530 - 540 - 550 - 560 - 570 - 580 - 590 - 600 - 610 - 620 - 630 - 640 - 650
660 - 670 - 680 - 690 - 700 - 710 - 720 - 730 - 740 - 750 - 760 - 770 - 780
790 - 800 - 810 - 820 - 830 - 840 - 850 - 860 - 870 - 880 - 890 - 900 - 910
920 - 930 - 940 - 950 - 960 - 970 - 980 - 990 - 1000

Notes: ______________________________

End Date: ____________

CLEANING

Holster Position:______________ **Hand:** ______ **Start Date:** _________
Spare Magazine Position: ___________ **Firearm:** __________________
Cover Garment: ________________________________ **Caliber:** ______
Notes:____

____

____

Slow Reps: 10 - 20 - 30 - 40 - 50 - 60 - 70 - 80 - 90 - 100 - 110 - 120 - 130
140 - 150 - 160 - 170 - 180 - 190 - 200 - 210 - 220 - 230 - 240 - 250 - 260
270 - 280 - 290 - 300 - 310 - 320 - 330 - 340 - 350 - 360 - 370 - 380 - 390
400 - 410 - 420 - 430 - 440 - 450 - 460 - 470 - 480 - 490 - 500 - 510 - 520
530 - 540 - 550 - 560 - 570 - 580 - 590 - 600 - 610 - 620 - 630 - 640 - 650
660 - 670 - 680 - 690 - 700 - 710 - 720 - 730 - 740 - 750 - 760 - 770 - 780
790 - 800 - 810 - 820 - 830 - 840 - 850 - 860 - 870 - 880 - 890 - 900 - 910
920 - 930 - 940 - 950 - 960 - 970 - 980 - 990 - 1000

Reg Reps: 10 - 20 - 30 - 40 - 50 - 60 - 70 - 80 - 90 - 100 - 110 - 120 - 130
140 - 150 - 160 - 170 - 180 - 190 - 200 - 210 - 220 - 230 - 240 - 250 - 260
270 - 280 - 290 - 300 - 310 - 320 - 330 - 340 - 350 - 360 - 370 - 380 - 390
400 - 410 - 420 - 430 - 440 - 450 - 460 - 470 - 480 - 490 - 500 - 510 - 520
530 - 540 - 550 - 560 - 570 - 580 - 590 - 600 - 610 - 620 - 630 - 640 - 650
660 - 670 - 680 - 690 - 700 - 710 - 720 - 730 - 740 - 750 - 760 - 770 - 780
790 - 800 - 810 - 820 - 830 - 840 - 850 - 860 - 870 - 880 - 890 - 900 - 910
920 - 930 - 940 - 950 - 960 - 970 - 980 - 990 - 1000
Notes: ____

____

____

Reg Reps: 10 - 20 - 30 - 40 - 50 - 60 - 70 - 80 - 90 - 100 - 110 - 120 - 130
140 - 150 - 160 - 170 - 180 - 190 - 200 - 210 - 220 - 230 - 240 - 250 - 260
270 - 280 - 290 - 300 - 310 - 320 - 330 - 340 - 350 - 360 - 370 - 380 - 390
400 - 410 - 420 - 430 - 440 - 450 - 460 - 470 - 480 - 490 - 500 - 510 - 520
530 - 540 - 550 - 560 - 570 - 580 - 590 - 600 - 610 - 620 - 630 - 640 - 650
660 - 670 - 680 - 690 - 700 - 710 - 720 - 730 - 740 - 750 - 760 - 770 - 780
790 - 800 - 810 - 820 - 830 - 840 - 850 - 860 - 870 - 880 - 890 - 900 - 910
920 - 930 - 940 - 950 - 960 - 970 - 980 - 990 - 1000
Notes: ____

____

____

Reg Reps: 10 - 20 - 30 - 40 - 50 - 60 - 70 - 80 - 90 - 100 - 110 - 120 - 130
140 - 150 - 160 - 170 - 180 - 190 - 200 - 210 - 220 - 230 - 240 - 250 - 260
270 - 280 - 290 - 300 - 310 - 320 - 330 - 340 - 350 - 360 - 370 - 380 - 390
400 - 410 - 420 - 430 - 440 - 450 - 460 - 470 - 480 - 490 - 500 - 510 - 520
530 - 540 - 550 - 560 - 570 - 580 - 590 - 600 - 610 - 620 - 630 - 640 - 650
660 - 670 - 680 - 690 - 700 - 710 - 720 - 730 - 740 - 750 - 760 - 770 - 780
790 - 800 - 810 - 820 - 830 - 840 - 850 - 860 - 870 - 880 - 890 - 900 - 910
920 - 930 - 940 - 950 - 960 - 970 - 980 - 990 - 1000

Notes: ____

____

____

____

Reg Reps: 10 - 20 - 30 - 40 - 50 - 60 - 70 - 80 - 90 - 100 - 110 - 120 - 130
140 - 150 - 160 - 170 - 180 - 190 - 200 - 210 - 220 - 230 - 240 - 250 - 260
270 - 280 - 290 - 300 - 310 - 320 - 330 - 340 - 350 - 360 - 370 - 380 - 390
400 - 410 - 420 - 430 - 440 - 450 - 460 - 470 - 480 - 490 - 500 - 510 - 520
530 - 540 - 550 - 560 - 570 - 580 - 590 - 600 - 610 - 620 - 630 - 640 - 650
660 - 670 - 680 - 690 - 700 - 710 - 720 - 730 - 740 - 750 - 760 - 770 - 780
790 - 800 - 810 - 820 - 830 - 840 - 850 - 860 - 870 - 880 - 890 - 900 - 910
920 - 930 - 940 - 950 - 960 - 970 - 980 - 990 - 1000

Notes: ____

____

____

____

Reg Reps: 10 - 20 - 30 - 40 - 50 - 60 - 70 - 80 - 90 - 100 - 110 - 120 - 130
140 - 150 - 160 - 170 - 180 - 190 - 200 - 210 - 220 - 230 - 240 - 250 - 260
270 - 280 - 290 - 300 - 310 - 320 - 330 - 340 - 350 - 360 - 370 - 380 - 390
400 - 410 - 420 - 430 - 440 - 450 - 460 - 470 - 480 - 490 - 500 - 510 - 520
530 - 540 - 550 - 560 - 570 - 580 - 590 - 600 - 610 - 620 - 630 - 640 - 650
660 - 670 - 680 - 690 - 700 - 710 - 720 - 730 - 740 - 750 - 760 - 770 - 780
790 - 800 - 810 - 820 - 830 - 840 - 850 - 860 - 870 - 880 - 890 - 900 - 910
920 - 930 - 940 - 950 - 960 - 970 - 980 - 990 - 1000

Notes: ____

____

____

______________________________ **End Date:** __________

GARMENT CLEARING

Holster Position:______________ **Hand:** ______ **Start Date:** __________
Spare Magazine Position: ____________ **Firearm:** ____________________
Cover Garment: __________________________________ **Caliber:** ______
Notes:____

____

____

Slow Reps: 10 - 20 - 30 - 40 - 50 - 60 - 70 - 80 - 90 - 100 - 110 - 120 - 130
140 - 150 - 160 - 170 - 180 - 190 - 200 - 210 - 220 - 230 - 240 - 250 - 260
270 - 280 - 290 - 300 - 310 - 320 - 330 - 340 - 350 - 360 - 370 - 380 - 390
400 - 410 - 420 - 430 - 440 - 450 - 460 - 470 - 480 - 490 - 500 - 510 - 520
530 - 540 - 550 - 560 - 570 - 580 - 590 - 600 - 610 - 620 - 630 - 640 - 650
660 - 670 - 680 - 690 - 700 - 710 - 720 - 730 - 740 - 750 - 760 - 770 - 780
790 - 800 - 810 - 820 - 830 - 840 - 850 - 860 - 870 - 880 - 890 - 900 - 910
920 - 930 - 940 - 950 - 960 - 970 - 980 - 990 - 1000

Reg Reps: 10 - 20 - 30 - 40 - 50 - 60 - 70 - 80 - 90 - 100 - 110 - 120 - 130
140 - 150 - 160 - 170 - 180 - 190 - 200 - 210 - 220 - 230 - 240 - 250 - 260
270 - 280 - 290 - 300 - 310 - 320 - 330 - 340 - 350 - 360 - 370 - 380 - 390
400 - 410 - 420 - 430 - 440 - 450 - 460 - 470 - 480 - 490 - 500 - 510 - 520
530 - 540 - 550 - 560 - 570 - 580 - 590 - 600 - 610 - 620 - 630 - 640 - 650
660 - 670 - 680 - 690 - 700 - 710 - 720 - 730 - 740 - 750 - 760 - 770 - 780
790 - 800 - 810 - 820 - 830 - 840 - 850 - 860 - 870 - 880 - 890 - 900 - 910
920 - 930 - 940 - 950 - 960 - 970 - 980 - 990 - 1000

Notes: ____

____

____

Reg Reps: 10 - 20 - 30 - 40 - 50 - 60 - 70 - 80 - 90 - 100 - 110 - 120 - 130
140 - 150 - 160 - 170 - 180 - 190 - 200 - 210 - 220 - 230 - 240 - 250 - 260
270 - 280 - 290 - 300 - 310 - 320 - 330 - 340 - 350 - 360 - 370 - 380 - 390
400 - 410 - 420 - 430 - 440 - 450 - 460 - 470 - 480 - 490 - 500 - 510 - 520
530 - 540 - 550 - 560 - 570 - 580 - 590 - 600 - 610 - 620 - 630 - 640 - 650
660 - 670 - 680 - 690 - 700 - 710 - 720 - 730 - 740 - 750 - 760 - 770 - 780
790 - 800 - 810 - 820 - 830 - 840 - 850 - 860 - 870 - 880 - 890 - 900 - 910
920 - 930 - 940 - 950 - 960 - 970 - 980 - 990 - 1000

Notes: ____

____

____

Reg Reps: 10 - 20 - 30 - 40 - 50 - 60 - 70 - 80 - 90 - 100 - 110 - 120 - 130
140 - 150 - 160 - 170 - 180 - 190 - 200 - 210 - 220 - 230 - 240 - 250 - 260
270 - 280 - 290 - 300 - 310 - 320 - 330 - 340 - 350 - 360 - 370 - 380 - 390
400 - 410 - 420 - 430 - 440 - 450 - 460 - 470 - 480 - 490 - 500 - 510 - 520
530 - 540 - 550 - 560 - 570 - 580 - 590 - 600 - 610 - 620 - 630 - 640 - 650
660 - 670 - 680 - 690 - 700 - 710 - 720 - 730 - 740 - 750 - 760 - 770 - 780
790 - 800 - 810 - 820 - 830 - 840 - 850 - 860 - 870 - 880 - 890 - 900 - 910
920 - 930 - 940 - 950 - 960 - 970 - 980 - 990 - 1000

Notes: ____

____

____

____

Reg Reps: 10 - 20 - 30 - 40 - 50 - 60 - 70 - 80 - 90 - 100 - 110 - 120 - 130
140 - 150 - 160 - 170 - 180 - 190 - 200 - 210 - 220 - 230 - 240 - 250 - 260
270 - 280 - 290 - 300 - 310 - 320 - 330 - 340 - 350 - 360 - 370 - 380 - 390
400 - 410 - 420 - 430 - 440 - 450 - 460 - 470 - 480 - 490 - 500 - 510 - 520
530 - 540 - 550 - 560 - 570 - 580 - 590 - 600 - 610 - 620 - 630 - 640 - 650
660 - 670 - 680 - 690 - 700 - 710 - 720 - 730 - 740 - 750 - 760 - 770 - 780
790 - 800 - 810 - 820 - 830 - 840 - 850 - 860 - 870 - 880 - 890 - 900 - 910
920 - 930 - 940 - 950 - 960 - 970 - 980 - 990 - 1000

Notes: ____

____

____

____

Reg Reps: 10 - 20 - 30 - 40 - 50 - 60 - 70 - 80 - 90 - 100 - 110 - 120 - 130
140 - 150 - 160 - 170 - 180 - 190 - 200 - 210 - 220 - 230 - 240 - 250 - 260
270 - 280 - 290 - 300 - 310 - 320 - 330 - 340 - 350 - 360 - 370 - 380 - 390
400 - 410 - 420 - 430 - 440 - 450 - 460 - 470 - 480 - 490 - 500 - 510 - 520
530 - 540 - 550 - 560 - 570 - 580 - 590 - 600 - 610 - 620 - 630 - 640 - 650
660 - 670 - 680 - 690 - 700 - 710 - 720 - 730 - 740 - 750 - 760 - 770 - 780
790 - 800 - 810 - 820 - 830 - 840 - 850 - 860 - 870 - 880 - 890 - 900 - 910
920 - 930 - 940 - 950 - 960 - 970 - 980 - 990 - 1000

Notes: ____

____

____

______________________________ **End Date:** __________

TWO-HANDED GRIP

Holster Position:_____________ **Hand:** ______ **Start Date:** _________
Spare Magazine Position: ___________ **Firearm:** __________________
Cover Garment: ________________________________ **Caliber:** ______
Notes:____
____
____

Slow Reps: 10 - 20 - 30 - 40 - 50 - 60 - 70 - 80 - 90 - 100 - 110 - 120 - 130
140 - 150 - 160 - 170 - 180 - 190 - 200 - 210 - 220 - 230 - 240 - 250 - 260
270 - 280 - 290 - 300 - 310 - 320 - 330 - 340 - 350 - 360 - 370 - 380 - 390
400 - 410 - 420 - 430 - 440 - 450 - 460 - 470 - 480 - 490 - 500 - 510 - 520
530 - 540 - 550 - 560 - 570 - 580 - 590 - 600 - 610 - 620 - 630 - 640 - 650
660 - 670 - 680 - 690 - 700 - 710 - 720 - 730 - 740 - 750 - 760 - 770 - 780
790 - 800 - 810 - 820 - 830 - 840 - 850 - 860 - 870 - 880 - 890 - 900 - 910
920 - 930 - 940 - 950 - 960 - 970 - 980 - 990 - 1000

Reg Reps: 10 - 20 - 30 - 40 - 50 - 60 - 70 - 80 - 90 - 100 - 110 - 120 - 130
140 - 150 - 160 - 170 - 180 - 190 - 200 - 210 - 220 - 230 - 240 - 250 - 260
270 - 280 - 290 - 300 - 310 - 320 - 330 - 340 - 350 - 360 - 370 - 380 - 390
400 - 410 - 420 - 430 - 440 - 450 - 460 - 470 - 480 - 490 - 500 - 510 - 520
530 - 540 - 550 - 560 - 570 - 580 - 590 - 600 - 610 - 620 - 630 - 640 - 650
660 - 670 - 680 - 690 - 700 - 710 - 720 - 730 - 740 - 750 - 760 - 770 - 780
790 - 800 - 810 - 820 - 830 - 840 - 850 - 860 - 870 - 880 - 890 - 900 - 910
920 - 930 - 940 - 950 - 960 - 970 - 980 - 990 - 1000
Notes: ____
____
____

Reg Reps: 10 - 20 - 30 - 40 - 50 - 60 - 70 - 80 - 90 - 100 - 110 - 120 - 130
140 - 150 - 160 - 170 - 180 - 190 - 200 - 210 - 220 - 230 - 240 - 250 - 260
270 - 280 - 290 - 300 - 310 - 320 - 330 - 340 - 350 - 360 - 370 - 380 - 390
400 - 410 - 420 - 430 - 440 - 450 - 460 - 470 - 480 - 490 - 500 - 510 - 520
530 - 540 - 550 - 560 - 570 - 580 - 590 - 600 - 610 - 620 - 630 - 640 - 650
660 - 670 - 680 - 690 - 700 - 710 - 720 - 730 - 740 - 750 - 760 - 770 - 780
790 - 800 - 810 - 820 - 830 - 840 - 850 - 860 - 870 - 880 - 890 - 900 - 910
920 - 930 - 940 - 950 - 960 - 970 - 980 - 990 - 1000
Notes: ____
____
____

Reg Reps: 10 - 20 - 30 - 40 - 50 - 60 - 70 - 80 - 90 - 100 - 110 - 120 - 130
140 - 150 - 160 - 170 - 180 - 190 - 200 - 210 - 220 - 230 - 240 - 250 - 260
270 - 280 - 290 - 300 - 310 - 320 - 330 - 340 - 350 - 360 - 370 - 380 - 390
400 - 410 - 420 - 430 - 440 - 450 - 460 - 470 - 480 - 490 - 500 - 510 - 520
530 - 540 - 550 - 560 - 570 - 580 - 590 - 600 - 610 - 620 - 630 - 640 - 650
660 - 670 - 680 - 690 - 700 - 710 - 720 - 730 - 740 - 750 - 760 - 770 - 780
790 - 800 - 810 - 820 - 830 - 840 - 850 - 860 - 870 - 880 - 890 - 900 - 910
920 - 930 - 940 - 950 - 960 - 970 - 980 - 990 - 1000

Notes: ____

____

____

____

Reg Reps: 10 - 20 - 30 - 40 - 50 - 60 - 70 - 80 - 90 - 100 - 110 - 120 - 130
140 - 150 - 160 - 170 - 180 - 190 - 200 - 210 - 220 - 230 - 240 - 250 - 260
270 - 280 - 290 - 300 - 310 - 320 - 330 - 340 - 350 - 360 - 370 - 380 - 390
400 - 410 - 420 - 430 - 440 - 450 - 460 - 470 - 480 - 490 - 500 - 510 - 520
530 - 540 - 550 - 560 - 570 - 580 - 590 - 600 - 610 - 620 - 630 - 640 - 650
660 - 670 - 680 - 690 - 700 - 710 - 720 - 730 - 740 - 750 - 760 - 770 - 780
790 - 800 - 810 - 820 - 830 - 840 - 850 - 860 - 870 - 880 - 890 - 900 - 910
920 - 930 - 940 - 950 - 960 - 970 - 980 - 990 - 1000

Notes: ____

____

____

____

Reg Reps: 10 - 20 - 30 - 40 - 50 - 60 - 70 - 80 - 90 - 100 - 110 - 120 - 130
140 - 150 - 160 - 170 - 180 - 190 - 200 - 210 - 220 - 230 - 240 - 250 - 260
270 - 280 - 290 - 300 - 310 - 320 - 330 - 340 - 350 - 360 - 370 - 380 - 390
400 - 410 - 420 - 430 - 440 - 450 - 460 - 470 - 480 - 490 - 500 - 510 - 520
530 - 540 - 550 - 560 - 570 - 580 - 590 - 600 - 610 - 620 - 630 - 640 - 650
660 - 670 - 680 - 690 - 700 - 710 - 720 - 730 - 740 - 750 - 760 - 770 - 780
790 - 800 - 810 - 820 - 830 - 840 - 850 - 860 - 870 - 880 - 890 - 900 - 910
920 - 930 - 940 - 950 - 960 - 970 - 980 - 990 - 1000

Notes: ____

____

____

______________________________ **End Date:** __________

Reg Reps: 10 - 20 - 30 - 40 - 50 - 60 - 70 - 80 - 90 - 100 - 110 - 120 - 130
140 - 150 - 160 - 170 - 180 - 190 - 200 - 210 - 220 - 230 - 240 - 250 - 260
270 - 280 - 290 - 300 - 310 - 320 - 330 - 340 - 350 - 360 - 370 - 380 - 390
400 - 410 - 420 - 430 - 440 - 450 - 460 - 470 - 480 - 490 - 500 - 510 - 520
530 - 540 - 550 - 560 - 570 - 580 - 590 - 600 - 610 - 620 - 630 - 640 - 650
660 - 670 - 680 - 690 - 700 - 710 - 720 - 730 - 740 - 750 - 760 - 770 - 780
790 - 800 - 810 - 820 - 830 - 840 - 850 - 860 - 870 - 880 - 890 - 900 - 910
920 - 930 - 940 - 950 - 960 - 970 - 980 - 990 - 1000

Notes: ______________________________

____

____

____

Reg Reps: 10 - 20 - 30 - 40 - 50 - 60 - 70 - 80 - 90 - 100 - 110 - 120 - 130
140 - 150 - 160 - 170 - 180 - 190 - 200 - 210 - 220 - 230 - 240 - 250 - 260
270 - 280 - 290 - 300 - 310 - 320 - 330 - 340 - 350 - 360 - 370 - 380 - 390
400 - 410 - 420 - 430 - 440 - 450 - 460 - 470 - 480 - 490 - 500 - 510 - 520
530 - 540 - 550 - 560 - 570 - 580 - 590 - 600 - 610 - 620 - 630 - 640 - 650
660 - 670 - 680 - 690 - 700 - 710 - 720 - 730 - 740 - 750 - 760 - 770 - 780
790 - 800 - 810 - 820 - 830 - 840 - 850 - 860 - 870 - 880 - 890 - 900 - 910
920 - 930 - 940 - 950 - 960 - 970 - 980 - 990 - 1000

Notes: ______________________________

____

____

____

Reg Reps: 10 - 20 - 30 - 40 - 50 - 60 - 70 - 80 - 90 - 100 - 110 - 120 - 130
140 - 150 - 160 - 170 - 180 - 190 - 200 - 210 - 220 - 230 - 240 - 250 - 260
270 - 280 - 290 - 300 - 310 - 320 - 330 - 340 - 350 - 360 - 370 - 380 - 390
400 - 410 - 420 - 430 - 440 - 450 - 460 - 470 - 480 - 490 - 500 - 510 - 520
530 - 540 - 550 - 560 - 570 - 580 - 590 - 600 - 610 - 620 - 630 - 640 - 650
660 - 670 - 680 - 690 - 700 - 710 - 720 - 730 - 740 - 750 - 760 - 770 - 780
790 - 800 - 810 - 820 - 830 - 840 - 850 - 860 - 870 - 880 - 890 - 900 - 910
920 - 930 - 940 - 950 - 960 - 970 - 980 - 990 - 1000

Notes: ______________________________

____

____

______________________________ **End Date:** __________

GARMENT CLEARING

Holster Position:_____________ **Hand:** ______ **Start Date:** _________
Spare Magazine Position: ___________ **Firearm:** ______________
Cover Garment: ______________________________ **Caliber:** ______
Notes:____
____
____

Slow Reps: 10 - 20 - 30 - 40 - 50 - 60 - 70 - 80 - 90 - 100 - 110 - 120 - 130
140 - 150 - 160 - 170 - 180 - 190 - 200 - 210 - 220 - 230 - 240 - 250 - 260
270 - 280 - 290 - 300 - 310 - 320 - 330 - 340 - 350 - 360 - 370 - 380 - 390
400 - 410 - 420 - 430 - 440 - 450 - 460 - 470 - 480 - 490 - 500 - 510 - 520
530 - 540 - 550 - 560 - 570 - 580 - 590 - 600 - 610 - 620 - 630 - 640 - 650
660 - 670 - 680 - 690 - 700 - 710 - 720 - 730 - 740 - 750 - 760 - 770 - 780
790 - 800 - 810 - 820 - 830 - 840 - 850 - 860 - 870 - 880 - 890 - 900 - 910
920 - 930 - 940 - 950 - 960 - 970 - 980 - 990 - 1000

Reg Reps: 10 - 20 - 30 - 40 - 50 - 60 - 70 - 80 - 90 - 100 - 110 - 120 - 130
140 - 150 - 160 - 170 - 180 - 190 - 200 - 210 - 220 - 230 - 240 - 250 - 260
270 - 280 - 290 - 300 - 310 - 320 - 330 - 340 - 350 - 360 - 370 - 380 - 390
400 - 410 - 420 - 430 - 440 - 450 - 460 - 470 - 480 - 490 - 500 - 510 - 520
530 - 540 - 550 - 560 - 570 - 580 - 590 - 600 - 610 - 620 - 630 - 640 - 650
660 - 670 - 680 - 690 - 700 - 710 - 720 - 730 - 740 - 750 - 760 - 770 - 780
790 - 800 - 810 - 820 - 830 - 840 - 850 - 860 - 870 - 880 - 890 - 900 - 910
920 - 930 - 940 - 950 - 960 - 970 - 980 - 990 - 1000
Notes: ____
____
____

Reg Reps: 10 - 20 - 30 - 40 - 50 - 60 - 70 - 80 - 90 - 100 - 110 - 120 - 130
140 - 150 - 160 - 170 - 180 - 190 - 200 - 210 - 220 - 230 - 240 - 250 - 260
270 - 280 - 290 - 300 - 310 - 320 - 330 - 340 - 350 - 360 - 370 - 380 - 390
400 - 410 - 420 - 430 - 440 - 450 - 460 - 470 - 480 - 490 - 500 - 510 - 520
530 - 540 - 550 - 560 - 570 - 580 - 590 - 600 - 610 - 620 - 630 - 640 - 650
660 - 670 - 680 - 690 - 700 - 710 - 720 - 730 - 740 - 750 - 760 - 770 - 780
790 - 800 - 810 - 820 - 830 - 840 - 850 - 860 - 870 - 880 - 890 - 900 - 910
920 - 930 - 940 - 950 - 960 - 970 - 980 - 990 - 1000
Notes: ____
____
____

MASTER GRIP

Holster Position: ____________ **Hand:** ______ **Start Date:** ________
Spare Magazine Position: __________ **Firearm:** ________________
Cover Garment: ______________________________ **Caliber:** ______
Notes: ____

____

____

Slow Reps: 10 - 20 - 30 - 40 - 50 - 60 - 70 - 80 - 90 - 100 - 110 - 120 - 130
140 - 150 - 160 - 170 - 180 - 190 - 200 - 210 - 220 - 230 - 240 - 250 - 260
270 - 280 - 290 - 300 - 310 - 320 - 330 - 340 - 350 - 360 - 370 - 380 - 390
400 - 410 - 420 - 430 - 440 - 450 - 460 - 470 - 480 - 490 - 500 - 510 - 520
530 - 540 - 550 - 560 - 570 - 580 - 590 - 600 - 610 - 620 - 630 - 640 - 650
660 - 670 - 680 - 690 - 700 - 710 - 720 - 730 - 740 - 750 - 760 - 770 - 780
790 - 800 - 810 - 820 - 830 - 840 - 850 - 860 - 870 - 880 - 890 - 900 - 910
920 - 930 - 940 - 950 - 960 - 970 - 980 - 990 - 1000

Reg Reps: 10 - 20 - 30 - 40 - 50 - 60 - 70 - 80 - 90 - 100 - 110 - 120 - 130
140 - 150 - 160 - 170 - 180 - 190 - 200 - 210 - 220 - 230 - 240 - 250 - 260
270 - 280 - 290 - 300 - 310 - 320 - 330 - 340 - 350 - 360 - 370 - 380 - 390
400 - 410 - 420 - 430 - 440 - 450 - 460 - 470 - 480 - 490 - 500 - 510 - 520
530 - 540 - 550 - 560 - 570 - 580 - 590 - 600 - 610 - 620 - 630 - 640 - 650
660 - 670 - 680 - 690 - 700 - 710 - 720 - 730 - 740 - 750 - 760 - 770 - 780
790 - 800 - 810 - 820 - 830 - 840 - 850 - 860 - 870 - 880 - 890 - 900 - 910
920 - 930 - 940 - 950 - 960 - 970 - 980 - 990 - 1000
Notes: ____

____

____

Reg Reps: 10 - 20 - 30 - 40 - 50 - 60 - 70 - 80 - 90 - 100 - 110 - 120 - 130
140 - 150 - 160 - 170 - 180 - 190 - 200 - 210 - 220 - 230 - 240 - 250 - 260
270 - 280 - 290 - 300 - 310 - 320 - 330 - 340 - 350 - 360 - 370 - 380 - 390
400 - 410 - 420 - 430 - 440 - 450 - 460 - 470 - 480 - 490 - 500 - 510 - 520
530 - 540 - 550 - 560 - 570 - 580 - 590 - 600 - 610 - 620 - 630 - 640 - 650
660 - 670 - 680 - 690 - 700 - 710 - 720 - 730 - 740 - 750 - 760 - 770 - 780
790 - 800 - 810 - 820 - 830 - 840 - 850 - 860 - 870 - 880 - 890 - 900 - 910
920 - 930 - 940 - 950 - 960 - 970 - 980 - 990 - 1000
Notes: ____

____

____

Reg Reps: 10 - 20 - 30 - 40 - 50 - 60 - 70 - 80 - 90 - 100 - 110 - 120 - 130
140 - 150 - 160 - 170 - 180 - 190 - 200 - 210 - 220 - 230 - 240 - 250 - 260
270 - 280 - 290 - 300 - 310 - 320 - 330 - 340 - 350 - 360 - 370 - 380 - 390
400 - 410 - 420 - 430 - 440 - 450 - 460 - 470 - 480 - 490 - 500 - 510 - 520
530 - 540 - 550 - 560 - 570 - 580 - 590 - 600 - 610 - 620 - 630 - 640 - 650
660 - 670 - 680 - 690 - 700 - 710 - 720 - 730 - 740 - 750 - 760 - 770 - 780
790 - 800 - 810 - 820 - 830 - 840 - 850 - 860 - 870 - 880 - 890 - 900 - 910
920 - 930 - 940 - 950 - 960 - 970 - 980 - 990 - 1000

Notes: ____

____

____

____

Reg Reps: 10 - 20 - 30 - 40 - 50 - 60 - 70 - 80 - 90 - 100 - 110 - 120 - 130
140 - 150 - 160 - 170 - 180 - 190 - 200 - 210 - 220 - 230 - 240 - 250 - 260
270 - 280 - 290 - 300 - 310 - 320 - 330 - 340 - 350 - 360 - 370 - 380 - 390
400 - 410 - 420 - 430 - 440 - 450 - 460 - 470 - 480 - 490 - 500 - 510 - 520
530 - 540 - 550 - 560 - 570 - 580 - 590 - 600 - 610 - 620 - 630 - 640 - 650
660 - 670 - 680 - 690 - 700 - 710 - 720 - 730 - 740 - 750 - 760 - 770 - 780
790 - 800 - 810 - 820 - 830 - 840 - 850 - 860 - 870 - 880 - 890 - 900 - 910
920 - 930 - 940 - 950 - 960 - 970 - 980 - 990 - 1000

Notes: ____

____

____

____

Reg Reps: 10 - 20 - 30 - 40 - 50 - 60 - 70 - 80 - 90 - 100 - 110 - 120 - 130
140 - 150 - 160 - 170 - 180 - 190 - 200 - 210 - 220 - 230 - 240 - 250 - 260
270 - 280 - 290 - 300 - 310 - 320 - 330 - 340 - 350 - 360 - 370 - 380 - 390
400 - 410 - 420 - 430 - 440 - 450 - 460 - 470 - 480 - 490 - 500 - 510 - 520
530 - 540 - 550 - 560 - 570 - 580 - 590 - 600 - 610 - 620 - 630 - 640 - 650
660 - 670 - 680 - 690 - 700 - 710 - 720 - 730 - 740 - 750 - 760 - 770 - 780
790 - 800 - 810 - 820 - 830 - 840 - 850 - 860 - 870 - 880 - 890 - 900 - 910
920 - 930 - 940 - 950 - 960 - 970 - 980 - 990 - 1000

Notes: ____

____

____

______________________________ **End Date:** __________

TRIGGER DISCIPLINE

Holster Position:_____________ **Hand:** ______ **Start Date:** ________

Spare Magazine Position: ___________ **Firearm:** ____________________

Cover Garment: ________________________________ **Caliber:** ______

Notes:____

____

____

Slow Reps: 10 - 20 - 30 - 40 - 50 - 60 - 70 - 80 - 90 - 100 - 110 - 120 - 130
140 - 150 - 160 - 170 - 180 - 190 - 200 - 210 - 220 - 230 - 240 - 250 - 260
270 - 280 - 290 - 300 - 310 - 320 - 330 - 340 - 350 - 360 - 370 - 380 - 390
400 - 410 - 420 - 430 - 440 - 450 - 460 - 470 - 480 - 490 - 500 - 510 - 520
530 - 540 - 550 - 560 - 570 - 580 - 590 - 600 - 610 - 620 - 630 - 640 - 650
660 - 670 - 680 - 690 - 700 - 710 - 720 - 730 - 740 - 750 - 760 - 770 - 780
790 - 800 - 810 - 820 - 830 - 840 - 850 - 860 - 870 - 880 - 890 - 900 - 910
920 - 930 - 940 - 950 - 960 - 970 - 980 - 990 - 1000

Reg Reps: 10 - 20 - 30 - 40 - 50 - 60 - 70 - 80 - 90 - 100 - 110 - 120 - 130
140 - 150 - 160 - 170 - 180 - 190 - 200 - 210 - 220 - 230 - 240 - 250 - 260
270 - 280 - 290 - 300 - 310 - 320 - 330 - 340 - 350 - 360 - 370 - 380 - 390
400 - 410 - 420 - 430 - 440 - 450 - 460 - 470 - 480 - 490 - 500 - 510 - 520
530 - 540 - 550 - 560 - 570 - 580 - 590 - 600 - 610 - 620 - 630 - 640 - 650
660 - 670 - 680 - 690 - 700 - 710 - 720 - 730 - 740 - 750 - 760 - 770 - 780
790 - 800 - 810 - 820 - 830 - 840 - 850 - 860 - 870 - 880 - 890 - 900 - 910
920 - 930 - 940 - 950 - 960 - 970 - 980 - 990 - 1000

Notes: ____

____

____

Reg Reps: 10 - 20 - 30 - 40 - 50 - 60 - 70 - 80 - 90 - 100 - 110 - 120 - 130
140 - 150 - 160 - 170 - 180 - 190 - 200 - 210 - 220 - 230 - 240 - 250 - 260
270 - 280 - 290 - 300 - 310 - 320 - 330 - 340 - 350 - 360 - 370 - 380 - 390
400 - 410 - 420 - 430 - 440 - 450 - 460 - 470 - 480 - 490 - 500 - 510 - 520
530 - 540 - 550 - 560 - 570 - 580 - 590 - 600 - 610 - 620 - 630 - 640 - 650
660 - 670 - 680 - 690 - 700 - 710 - 720 - 730 - 740 - 750 - 760 - 770 - 780
790 - 800 - 810 - 820 - 830 - 840 - 850 - 860 - 870 - 880 - 890 - 900 - 910
920 - 930 - 940 - 950 - 960 - 970 - 980 - 990 - 1000

Notes: ____

____

____

Reg Reps: 10 - 20 - 30 - 40 - 50 - 60 - 70 - 80 - 90 - 100 - 110 - 120 - 130
140 - 150 - 160 - 170 - 180 - 190 - 200 - 210 - 220 - 230 - 240 - 250 - 260
270 - 280 - 290 - 300 - 310 - 320 - 330 - 340 - 350 - 360 - 370 - 380 - 390
400 - 410 - 420 - 430 - 440 - 450 - 460 - 470 - 480 - 490 - 500 - 510 - 520
530 - 540 - 550 - 560 - 570 - 580 - 590 - 600 - 610 - 620 - 630 - 640 - 650
660 - 670 - 680 - 690 - 700 - 710 - 720 - 730 - 740 - 750 - 760 - 770 - 780
790 - 800 - 810 - 820 - 830 - 840 - 850 - 860 - 870 - 880 - 890 - 900 - 910
920 - 930 - 940 - 950 - 960 - 970 - 980 - 990 - 1000

Notes: ____________________

Reg Reps: 10 - 20 - 30 - 40 - 50 - 60 - 70 - 80 - 90 - 100 - 110 - 120 - 130
140 - 150 - 160 - 170 - 180 - 190 - 200 - 210 - 220 - 230 - 240 - 250 - 260
270 - 280 - 290 - 300 - 310 - 320 - 330 - 340 - 350 - 360 - 370 - 380 - 390
400 - 410 - 420 - 430 - 440 - 450 - 460 - 470 - 480 - 490 - 500 - 510 - 520
530 - 540 - 550 - 560 - 570 - 580 - 590 - 600 - 610 - 620 - 630 - 640 - 650
660 - 670 - 680 - 690 - 700 - 710 - 720 - 730 - 740 - 750 - 760 - 770 - 780
790 - 800 - 810 - 820 - 830 - 840 - 850 - 860 - 870 - 880 - 890 - 900 - 910
920 - 930 - 940 - 950 - 960 - 970 - 980 - 990 - 1000

Notes: ____________________

Reg Reps: 10 - 20 - 30 - 40 - 50 - 60 - 70 - 80 - 90 - 100 - 110 - 120 - 130
140 - 150 - 160 - 170 - 180 - 190 - 200 - 210 - 220 - 230 - 240 - 250 - 260
270 - 280 - 290 - 300 - 310 - 320 - 330 - 340 - 350 - 360 - 370 - 380 - 390
400 - 410 - 420 - 430 - 440 - 450 - 460 - 470 - 480 - 490 - 500 - 510 - 520
530 - 540 - 550 - 560 - 570 - 580 - 590 - 600 - 610 - 620 - 630 - 640 - 650
660 - 670 - 680 - 690 - 700 - 710 - 720 - 730 - 740 - 750 - 760 - 770 - 780
790 - 800 - 810 - 820 - 830 - 840 - 850 - 860 - 870 - 880 - 890 - 900 - 910
920 - 930 - 940 - 950 - 960 - 970 - 980 - 990 - 1000

Notes: ____________________

End Date: __________

SIGHT PICTURE/SIGHT ALIGNMENT

Holster Position: ____________ **Hand:** ______ **Start Date:** _________
Spare Magazine Position: ___________ **Firearm:** __________________
Cover Garment: _______________________________ **Caliber:** ______
Notes: ____
____
____

Slow Reps: 10 - 20 - 30 - 40 - 50 - 60 - 70 - 80 - 90 - 100 - 110 - 120 - 130
140 - 150 - 160 - 170 - 180 - 190 - 200 - 210 - 220 - 230 - 240 - 250 - 260
270 - 280 - 290 - 300 - 310 - 320 - 330 - 340 - 350 - 360 - 370 - 380 - 390
400 - 410 - 420 - 430 - 440 - 450 - 460 - 470 - 480 - 490 - 500 - 510 - 520
530 - 540 - 550 - 560 - 570 - 580 - 590 - 600 - 610 - 620 - 630 - 640 - 650
660 - 670 - 680 - 690 - 700 - 710 - 720 - 730 - 740 - 750 - 760 - 770 - 780
790 - 800 - 810 - 820 - 830 - 840 - 850 - 860 - 870 - 880 - 890 - 900 - 910
920 - 930 - 940 - 950 - 960 - 970 - 980 - 990 - 1000

Reg Reps: 10 - 20 - 30 - 40 - 50 - 60 - 70 - 80 - 90 - 100 - 110 - 120 - 130
140 - 150 - 160 - 170 - 180 - 190 - 200 - 210 - 220 - 230 - 240 - 250 - 260
270 - 280 - 290 - 300 - 310 - 320 - 330 - 340 - 350 - 360 - 370 - 380 - 390
400 - 410 - 420 - 430 - 440 - 450 - 460 - 470 - 480 - 490 - 500 - 510 - 520
530 - 540 - 550 - 560 - 570 - 580 - 590 - 600 - 610 - 620 - 630 - 640 - 650
660 - 670 - 680 - 690 - 700 - 710 - 720 - 730 - 740 - 750 - 760 - 770 - 780
790 - 800 - 810 - 820 - 830 - 840 - 850 - 860 - 870 - 880 - 890 - 900 - 910
920 - 930 - 940 - 950 - 960 - 970 - 980 - 990 - 1000
Notes: ____
____
____

Reg Reps: 10 - 20 - 30 - 40 - 50 - 60 - 70 - 80 - 90 - 100 - 110 - 120 - 130
140 - 150 - 160 - 170 - 180 - 190 - 200 - 210 - 220 - 230 - 240 - 250 - 260
270 - 280 - 290 - 300 - 310 - 320 - 330 - 340 - 350 - 360 - 370 - 380 - 390
400 - 410 - 420 - 430 - 440 - 450 - 460 - 470 - 480 - 490 - 500 - 510 - 520
530 - 540 - 550 - 560 - 570 - 580 - 590 - 600 - 610 - 620 - 630 - 640 - 650
660 - 670 - 680 - 690 - 700 - 710 - 720 - 730 - 740 - 750 - 760 - 770 - 780
790 - 800 - 810 - 820 - 830 - 840 - 850 - 860 - 870 - 880 - 890 - 900 - 910
920 - 930 - 940 - 950 - 960 - 970 - 980 - 990 - 1000
Notes: ____
____
____

Reg Reps: 10 - 20 - 30 - 40 - 50 - 60 - 70 - 80 - 90 - 100 - 110 - 120 - 130 140 - 150 - 160 - 170 - 180 - 190 - 200 - 210 - 220 - 230 - 240 - 250 - 260 270 - 280 - 290 - 300 - 310 - 320 - 330 - 340 - 350 - 360 - 370 - 380 - 390 400 - 410 - 420 - 430 - 440 - 450 - 460 - 470 - 480 - 490 - 500 - 510 - 520 530 - 540 - 550 - 560 - 570 - 580 - 590 - 600 - 610 - 620 - 630 - 640 - 650 660 - 670 - 680 - 690 - 700 - 710 - 720 - 730 - 740 - 750 - 760 - 770 - 780 790 - 800 - 810 - 820 - 830 - 840 - 850 - 860 - 870 - 880 - 890 - 900 - 910 920 - 930 - 940 - 950 - 960 - 970 - 980 - 990 - 1000

Notes: ______________________________

____

____

____

Reg Reps: 10 - 20 - 30 - 40 - 50 - 60 - 70 - 80 - 90 - 100 - 110 - 120 - 130 140 - 150 - 160 - 170 - 180 - 190 - 200 - 210 - 220 - 230 - 240 - 250 - 260 270 - 280 - 290 - 300 - 310 - 320 - 330 - 340 - 350 - 360 - 370 - 380 - 390 400 - 410 - 420 - 430 - 440 - 450 - 460 - 470 - 480 - 490 - 500 - 510 - 520 530 - 540 - 550 - 560 - 570 - 580 - 590 - 600 - 610 - 620 - 630 - 640 - 650 660 - 670 - 680 - 690 - 700 - 710 - 720 - 730 - 740 - 750 - 760 - 770 - 780 790 - 800 - 810 - 820 - 830 - 840 - 850 - 860 - 870 - 880 - 890 - 900 - 910 920 - 930 - 940 - 950 - 960 - 970 - 980 - 990 - 1000

Notes: ______________________________

____

____

____

Reg Reps: 10 - 20 - 30 - 40 - 50 - 60 - 70 - 80 - 90 - 100 - 110 - 120 - 130 140 - 150 - 160 - 170 - 180 - 190 - 200 - 210 - 220 - 230 - 240 - 250 - 260 270 - 280 - 290 - 300 - 310 - 320 - 330 - 340 - 350 - 360 - 370 - 380 - 390 400 - 410 - 420 - 430 - 440 - 450 - 460 - 470 - 480 - 490 - 500 - 510 - 520 530 - 540 - 550 - 560 - 570 - 580 - 590 - 600 - 610 - 620 - 630 - 640 - 650 660 - 670 - 680 - 690 - 700 - 710 - 720 - 730 - 740 - 750 - 760 - 770 - 780 790 - 800 - 810 - 820 - 830 - 840 - 850 - 860 - 870 - 880 - 890 - 900 - 910 920 - 930 - 940 - 950 - 960 - 970 - 980 - 990 - 1000

Notes: ______________________________

____

____

______________________________ **End Date:** __________

REHOLSTERING

Holster Position:______________ **Hand:** ______ **Start Date:** _________
Spare Magazine Position: ___________ **Firearm:** ________________
Cover Garment: ________________________________ **Caliber:** ______
Notes:____

____

____

Slow Reps: 10 - 20 - 30 - 40 - 50 - 60 - 70 - 80 - 90 - 100 - 110 - 120 - 130
140 - 150 - 160 - 170 - 180 - 190 - 200 - 210 - 220 - 230 - 240 - 250 - 260
270 - 280 - 290 - 300 - 310 - 320 - 330 - 340 - 350 - 360 - 370 - 380 - 390
400 - 410 - 420 - 430 - 440 - 450 - 460 - 470 - 480 - 490 - 500 - 510 - 520
530 - 540 - 550 - 560 - 570 - 580 - 590 - 600 - 610 - 620 - 630 - 640 - 650
660 - 670 - 680 - 690 - 700 - 710 - 720 - 730 - 740 - 750 - 760 - 770 - 780
790 - 800 - 810 - 820 - 830 - 840 - 850 - 860 - 870 - 880 - 890 - 900 - 910
920 - 930 - 940 - 950 - 960 - 970 - 980 - 990 - 1000

Reg Reps: 10 - 20 - 30 - 40 - 50 - 60 - 70 - 80 - 90 - 100 - 110 - 120 - 130
140 - 150 - 160 - 170 - 180 - 190 - 200 - 210 - 220 - 230 - 240 - 250 - 260
270 - 280 - 290 - 300 - 310 - 320 - 330 - 340 - 350 - 360 - 370 - 380 - 390
400 - 410 - 420 - 430 - 440 - 450 - 460 - 470 - 480 - 490 - 500 - 510 - 520
530 - 540 - 550 - 560 - 570 - 580 - 590 - 600 - 610 - 620 - 630 - 640 - 650
660 - 670 - 680 - 690 - 700 - 710 - 720 - 730 - 740 - 750 - 760 - 770 - 780
790 - 800 - 810 - 820 - 830 - 840 - 850 - 860 - 870 - 880 - 890 - 900 - 910
920 - 930 - 940 - 950 - 960 - 970 - 980 - 990 - 1000

Notes: ____

____

____

Reg Reps: 10 - 20 - 30 - 40 - 50 - 60 - 70 - 80 - 90 - 100 - 110 - 120 - 130
140 - 150 - 160 - 170 - 180 - 190 - 200 - 210 - 220 - 230 - 240 - 250 - 260
270 - 280 - 290 - 300 - 310 - 320 - 330 - 340 - 350 - 360 - 370 - 380 - 390
400 - 410 - 420 - 430 - 440 - 450 - 460 - 470 - 480 - 490 - 500 - 510 - 520
530 - 540 - 550 - 560 - 570 - 580 - 590 - 600 - 610 - 620 - 630 - 640 - 650
660 - 670 - 680 - 690 - 700 - 710 - 720 - 730 - 740 - 750 - 760 - 770 - 780
790 - 800 - 810 - 820 - 830 - 840 - 850 - 860 - 870 - 880 - 890 - 900 - 910
920 - 930 - 940 - 950 - 960 - 970 - 980 - 990 - 1000

Notes: ____

____

____

Reg Reps: 10 - 20 - 30 - 40 - 50 - 60 - 70 - 80 - 90 - 100 - 110 - 120 - 130
140 - 150 - 160 - 170 - 180 - 190 - 200 - 210 - 220 - 230 - 240 - 250 - 260
270 - 280 - 290 - 300 - 310 - 320 - 330 - 340 - 350 - 360 - 370 - 380 - 390
400 - 410 - 420 - 430 - 440 - 450 - 460 - 470 - 480 - 490 - 500 - 510 - 520
530 - 540 - 550 - 560 - 570 - 580 - 590 - 600 - 610 - 620 - 630 - 640 - 650
660 - 670 - 680 - 690 - 700 - 710 - 720 - 730 - 740 - 750 - 760 - 770 - 780
790 - 800 - 810 - 820 - 830 - 840 - 850 - 860 - 870 - 880 - 890 - 900 - 910
920 - 930 - 940 - 950 - 960 - 970 - 980 - 990 - 1000

Notes: ____

____

____

____

Reg Reps: 10 - 20 - 30 - 40 - 50 - 60 - 70 - 80 - 90 - 100 - 110 - 120 - 130
140 - 150 - 160 - 170 - 180 - 190 - 200 - 210 - 220 - 230 - 240 - 250 - 260
270 - 280 - 290 - 300 - 310 - 320 - 330 - 340 - 350 - 360 - 370 - 380 - 390
400 - 410 - 420 - 430 - 440 - 450 - 460 - 470 - 480 - 490 - 500 - 510 - 520
530 - 540 - 550 - 560 - 570 - 580 - 590 - 600 - 610 - 620 - 630 - 640 - 650
660 - 670 - 680 - 690 - 700 - 710 - 720 - 730 - 740 - 750 - 760 - 770 - 780
790 - 800 - 810 - 820 - 830 - 840 - 850 - 860 - 870 - 880 - 890 - 900 - 910
920 - 930 - 940 - 950 - 960 - 970 - 980 - 990 - 1000

Notes: ____

____

____

____

Reg Reps: 10 - 20 - 30 - 40 - 50 - 60 - 70 - 80 - 90 - 100 - 110 - 120 - 130
140 - 150 - 160 - 170 - 180 - 190 - 200 - 210 - 220 - 230 - 240 - 250 - 260
270 - 280 - 290 - 300 - 310 - 320 - 330 - 340 - 350 - 360 - 370 - 380 - 390
400 - 410 - 420 - 430 - 440 - 450 - 460 - 470 - 480 - 490 - 500 - 510 - 520
530 - 540 - 550 - 560 - 570 - 580 - 590 - 600 - 610 - 620 - 630 - 640 - 650
660 - 670 - 680 - 690 - 700 - 710 - 720 - 730 - 740 - 750 - 760 - 770 - 780
790 - 800 - 810 - 820 - 830 - 840 - 850 - 860 - 870 - 880 - 890 - 900 - 910
920 - 930 - 940 - 950 - 960 - 970 - 980 - 990 - 1000

Notes: ____

____

____

______________________________ **End Date:** ____________

DEPRESSING THE TRIGGER

Holster Position:______________ **Hand:** ______ **Start Date:** _________
Spare Magazine Position: ____________ **Firearm:** ___________________
Cover Garment: __________________________________ **Caliber:** ______
Notes:____

____

____

Slow Reps: 10 - 20 - 30 - 40 - 50 - 60 - 70 - 80 - 90 - 100 - 110 - 120 - 130
140 - 150 - 160 - 170 - 180 - 190 - 200 - 210 - 220 - 230 - 240 - 250 - 260
270 - 280 - 290 - 300 - 310 - 320 - 330 - 340 - 350 - 360 - 370 - 380 - 390
400 - 410 - 420 - 430 - 440 - 450 - 460 - 470 - 480 - 490 - 500 - 510 - 520
530 - 540 - 550 - 560 - 570 - 580 - 590 - 600 - 610 - 620 - 630 - 640 - 650
660 - 670 - 680 - 690 - 700 - 710 - 720 - 730 - 740 - 750 - 760 - 770 - 780
790 - 800 - 810 - 820 - 830 - 840 - 850 - 860 - 870 - 880 - 890 - 900 - 910
920 - 930 - 940 - 950 - 960 - 970 - 980 - 990 - 1000

Reg Reps: 10 - 20 - 30 - 40 - 50 - 60 - 70 - 80 - 90 - 100 - 110 - 120 - 130
140 - 150 - 160 - 170 - 180 - 190 - 200 - 210 - 220 - 230 - 240 - 250 - 260
270 - 280 - 290 - 300 - 310 - 320 - 330 - 340 - 350 - 360 - 370 - 380 - 390
400 - 410 - 420 - 430 - 440 - 450 - 460 - 470 - 480 - 490 - 500 - 510 - 520
530 - 540 - 550 - 560 - 570 - 580 - 590 - 600 - 610 - 620 - 630 - 640 - 650
660 - 670 - 680 - 690 - 700 - 710 - 720 - 730 - 740 - 750 - 760 - 770 - 780
790 - 800 - 810 - 820 - 830 - 840 - 850 - 860 - 870 - 880 - 890 - 900 - 910
920 - 930 - 940 - 950 - 960 - 970 - 980 - 990 - 1000
Notes: ____

____

____

Reg Reps: 10 - 20 - 30 - 40 - 50 - 60 - 70 - 80 - 90 - 100 - 110 - 120 - 130
140 - 150 - 160 - 170 - 180 - 190 - 200 - 210 - 220 - 230 - 240 - 250 - 260
270 - 280 - 290 - 300 - 310 - 320 - 330 - 340 - 350 - 360 - 370 - 380 - 390
400 - 410 - 420 - 430 - 440 - 450 - 460 - 470 - 480 - 490 - 500 - 510 - 520
530 - 540 - 550 - 560 - 570 - 580 - 590 - 600 - 610 - 620 - 630 - 640 - 650
660 - 670 - 680 - 690 - 700 - 710 - 720 - 730 - 740 - 750 - 760 - 770 - 780
790 - 800 - 810 - 820 - 830 - 840 - 850 - 860 - 870 - 880 - 890 - 900 - 910
920 - 930 - 940 - 950 - 960 - 970 - 980 - 990 - 1000
Notes: ____

____

____

Reg Reps: 10 - 20 - 30 - 40 - 50 - 60 - 70 - 80 - 90 - 100 - 110 - 120 - 130
140 - 150 - 160 - 170 - 180 - 190 - 200 - 210 - 220 - 230 - 240 - 250 - 260
270 - 280 - 290 - 300 - 310 - 320 - 330 - 340 - 350 - 360 - 370 - 380 - 390
400 - 410 - 420 - 430 - 440 - 450 - 460 - 470 - 480 - 490 - 500 - 510 - 520
530 - 540 - 550 - 560 - 570 - 580 - 590 - 600 - 610 - 620 - 630 - 640 - 650
660 - 670 - 680 - 690 - 700 - 710 - 720 - 730 - 740 - 750 - 760 - 770 - 780
790 - 800 - 810 - 820 - 830 - 840 - 850 - 860 - 870 - 880 - 890 - 900 - 910
920 - 930 - 940 - 950 - 960 - 970 - 980 - 990 - 1000

Notes: ____________________

Reg Reps: 10 - 20 - 30 - 40 - 50 - 60 - 70 - 80 - 90 - 100 - 110 - 120 - 130
140 - 150 - 160 - 170 - 180 - 190 - 200 - 210 - 220 - 230 - 240 - 250 - 260
270 - 280 - 290 - 300 - 310 - 320 - 330 - 340 - 350 - 360 - 370 - 380 - 390
400 - 410 - 420 - 430 - 440 - 450 - 460 - 470 - 480 - 490 - 500 - 510 - 520
530 - 540 - 550 - 560 - 570 - 580 - 590 - 600 - 610 - 620 - 630 - 640 - 650
660 - 670 - 680 - 690 - 700 - 710 - 720 - 730 - 740 - 750 - 760 - 770 - 780
790 - 800 - 810 - 820 - 830 - 840 - 850 - 860 - 870 - 880 - 890 - 900 - 910
920 - 930 - 940 - 950 - 960 - 970 - 980 - 990 - 1000

Notes: ____________________

Reg Reps: 10 - 20 - 30 - 40 - 50 - 60 - 70 - 80 - 90 - 100 - 110 - 120 - 130
140 - 150 - 160 - 170 - 180 - 190 - 200 - 210 - 220 - 230 - 240 - 250 - 260
270 - 280 - 290 - 300 - 310 - 320 - 330 - 340 - 350 - 360 - 370 - 380 - 390
400 - 410 - 420 - 430 - 440 - 450 - 460 - 470 - 480 - 490 - 500 - 510 - 520
530 - 540 - 550 - 560 - 570 - 580 - 590 - 600 - 610 - 620 - 630 - 640 - 650
660 - 670 - 680 - 690 - 700 - 710 - 720 - 730 - 740 - 750 - 760 - 770 - 780
790 - 800 - 810 - 820 - 830 - 840 - 850 - 860 - 870 - 880 - 890 - 900 - 910
920 - 930 - 940 - 950 - 960 - 970 - 980 - 990 - 1000

Notes: ____________________

End Date: __________

MALFUNCTION DRILL #1: MISFIRE

Holster Position:__________ **Hand:** _____ **Start Date:** ________
Spare Magazine Position: __________ **Firearm:** ______________
Cover Garment: ____________________________ **Caliber:** _____
Notes:____

____

____

Slow Reps: 10 - 20 - 30 - 40 - 50 - 60 - 70 - 80 - 90 - 100 - 110 - 120 - 130
140 - 150 - 160 - 170 - 180 - 190 - 200 - 210 - 220 - 230 - 240 - 250 - 260
270 - 280 - 290 - 300 - 310 - 320 - 330 - 340 - 350 - 360 - 370 - 380 - 390
400 - 410 - 420 - 430 - 440 - 450 - 460 - 470 - 480 - 490 - 500 - 510 - 520
530 - 540 - 550 - 560 - 570 - 580 - 590 - 600 - 610 - 620 - 630 - 640 - 650
660 - 670 - 680 - 690 - 700 - 710 - 720 - 730 - 740 - 750 - 760 - 770 - 780
790 - 800 - 810 - 820 - 830 - 840 - 850 - 860 - 870 - 880 - 890 - 900 - 910
920 - 930 - 940 - 950 - 960 - 970 - 980 - 990 - 1000

Reg Reps: 10 - 20 - 30 - 40 - 50 - 60 - 70 - 80 - 90 - 100 - 110 - 120 - 130
140 - 150 - 160 - 170 - 180 - 190 - 200 - 210 - 220 - 230 - 240 - 250 - 260
270 - 280 - 290 - 300 - 310 - 320 - 330 - 340 - 350 - 360 - 370 - 380 - 390
400 - 410 - 420 - 430 - 440 - 450 - 460 - 470 - 480 - 490 - 500 - 510 - 520
530 - 540 - 550 - 560 - 570 - 580 - 590 - 600 - 610 - 620 - 630 - 640 - 650
660 - 670 - 680 - 690 - 700 - 710 - 720 - 730 - 740 - 750 - 760 - 770 - 780
790 - 800 - 810 - 820 - 830 - 840 - 850 - 860 - 870 - 880 - 890 - 900 - 910
920 - 930 - 940 - 950 - 960 - 970 - 980 - 990 - 1000
Notes: ____

____

____

Reg Reps: 10 - 20 - 30 - 40 - 50 - 60 - 70 - 80 - 90 - 100 - 110 - 120 - 130
140 - 150 - 160 - 170 - 180 - 190 - 200 - 210 - 220 - 230 - 240 - 250 - 260
270 - 280 - 290 - 300 - 310 - 320 - 330 - 340 - 350 - 360 - 370 - 380 - 390
400 - 410 - 420 - 430 - 440 - 450 - 460 - 470 - 480 - 490 - 500 - 510 - 520
530 - 540 - 550 - 560 - 570 - 580 - 590 - 600 - 610 - 620 - 630 - 640 - 650
660 - 670 - 680 - 690 - 700 - 710 - 720 - 730 - 740 - 750 - 760 - 770 - 780
790 - 800 - 810 - 820 - 830 - 840 - 850 - 860 - 870 - 880 - 890 - 900 - 910
920 - 930 - 940 - 950 - 960 - 970 - 980 - 990 - 1000
Notes: ____

____

____

Reg Reps: 10 - 20 - 30 - 40 - 50 - 60 - 70 - 80 - 90 - 100 - 110 - 120 - 130 140 - 150 - 160 - 170 - 180 - 190 - 200 - 210 - 220 - 230 - 240 - 250 - 260 270 - 280 - 290 - 300 - 310 - 320 - 330 - 340 - 350 - 360 - 370 - 380 - 390 400 - 410 - 420 - 430 - 440 - 450 - 460 - 470 - 480 - 490 - 500 - 510 - 520 530 - 540 - 550 - 560 - 570 - 580 - 590 - 600 - 610 - 620 - 630 - 640 - 650 660 - 670 - 680 - 690 - 700 - 710 - 720 - 730 - 740 - 750 - 760 - 770 - 780 790 - 800 - 810 - 820 - 830 - 840 - 850 - 860 - 870 - 880 - 890 - 900 - 910 920 - 930 - 940 - 950 - 960 - 970 - 980 - 990 - 1000

Notes: ____

____

____

____

Reg Reps: 10 - 20 - 30 - 40 - 50 - 60 - 70 - 80 - 90 - 100 - 110 - 120 - 130 140 - 150 - 160 - 170 - 180 - 190 - 200 - 210 - 220 - 230 - 240 - 250 - 260 270 - 280 - 290 - 300 - 310 - 320 - 330 - 340 - 350 - 360 - 370 - 380 - 390 400 - 410 - 420 - 430 - 440 - 450 - 460 - 470 - 480 - 490 - 500 - 510 - 520 530 - 540 - 550 - 560 - 570 - 580 - 590 - 600 - 610 - 620 - 630 - 640 - 650 660 - 670 - 680 - 690 - 700 - 710 - 720 - 730 - 740 - 750 - 760 - 770 - 780 790 - 800 - 810 - 820 - 830 - 840 - 850 - 860 - 870 - 880 - 890 - 900 - 910 920 - 930 - 940 - 950 - 960 - 970 - 980 - 990 - 1000

Notes: ____

____

____

____

Reg Reps: 10 - 20 - 30 - 40 - 50 - 60 - 70 - 80 - 90 - 100 - 110 - 120 - 130 140 - 150 - 160 - 170 - 180 - 190 - 200 - 210 - 220 - 230 - 240 - 250 - 260 270 - 280 - 290 - 300 - 310 - 320 - 330 - 340 - 350 - 360 - 370 - 380 - 390 400 - 410 - 420 - 430 - 440 - 450 - 460 - 470 - 480 - 490 - 500 - 510 - 520 530 - 540 - 550 - 560 - 570 - 580 - 590 - 600 - 610 - 620 - 630 - 640 - 650 660 - 670 - 680 - 690 - 700 - 710 - 720 - 730 - 740 - 750 - 760 - 770 - 780 790 - 800 - 810 - 820 - 830 - 840 - 850 - 860 - 870 - 880 - 890 - 900 - 910 920 - 930 - 940 - 950 - 960 - 970 - 980 - 990 - 1000

Notes: ____

____

____

______________________________ **End Date:** ____________

s: 10 - 20 - 30 - 40 - 50 - 60 - 70 - 80 - 90 - 100 - 110 - 120 - 130
50 - 160 - 170 - 180 - 190 - 200 - 210 - 220 - 230 - 240 - 250 - 260
80 - 290 - 300 - 310 - 320 - 330 - 340 - 350 - 360 - 370 - 380 - 390
10 - 420 - 430 - 440 - 450 - 460 - 470 - 480 - 490 - 500 - 510 - 520
40 - 550 - 560 - 570 - 580 - 590 - 600 - 610 - 620 - 630 - 640 - 650
70 - 680 - 690 - 700 - 710 - 720 - 730 - 740 - 750 - 760 - 770 - 780
00 - 810 - 820 - 830 - 840 - 850 - 860 - 870 - 880 - 890 - 900 - 910
30 - 940 - 950 - 960 - 970 - 980 - 990 - 1000

s: 10 - 20 - 30 - 40 - 50 - 60 - 70 - 80 - 90 - 100 - 110 - 120 - 130
50 - 160 - 170 - 180 - 190 - 200 - 210 - 220 - 230 - 240 - 250 - 260
80 - 290 - 300 - 310 - 320 - 330 - 340 - 350 - 360 - 370 - 380 - 390
10 - 420 - 430 - 440 - 450 - 460 - 470 - 480 - 490 - 500 - 510 - 520
40 - 550 - 560 - 570 - 580 - 590 - 600 - 610 - 620 - 630 - 640 - 650
70 - 680 - 690 - 700 - 710 - 720 - 730 - 740 - 750 - 760 - 770 - 780
00 - 810 - 820 - 830 - 840 - 850 - 860 - 870 - 880 - 890 - 900 - 910
30 - 940 - 950 - 960 - 970 - 980 - 990 - 1000

s: 10 - 20 - 30 - 40 - 50 - 60 - 70 - 80 - 90 - 100 - 110 - 120 - 130
50 - 160 - 170 - 180 - 190 - 200 - 210 - 220 - 230 - 240 - 250 - 260
80 - 290 - 300 - 310 - 320 - 330 - 340 - 350 - 360 - 370 - 380 - 390
10 - 420 - 430 - 440 - 450 - 460 - 470 - 480 - 490 - 500 - 510 - 520
40 - 550 - 560 - 570 - 580 - 590 - 600 - 610 - 620 - 630 - 640 - 650
70 - 680 - 690 - 700 - 710 - 720 - 730 - 740 - 750 - 760 - 770 - 780
00 - 810 - 820 - 830 - 840 - 850 - 860 - 870 - 880 - 890 - 900 - 910
30 - 940 - 950 - 960 - 970 - 980 - 990 - 1000

End Date:

MALFUNCTION DRILL #2: ST

Holster Position:____________ **Hand:** ______
Sparc Magazinc Position: ___________ **Firearı**
Cover Garment: ________________________
Notes:________________________________

Slow Reps: 10 - 20 - 30 - 40 - 50 - 60 - 70 - 80 - 9
140 - 150 - 160 - 170 - 180 - 190 - 200 - 210 - 2
270 - 280 - 290 - 300 - 310 - 320 - 330 - 340 - 3
400 - 410 - 420 - 430 - 440 - 450 - 460 - 470 - 4
530 - 540 - 550 - 560 - 570 - 580 - 590 - 600 - 6
660 - 670 - 680 - 690 - 700 - 710 - 720 - 730 - 7
790 - 800 - 810 - 820 - 830 - 840 - 850 - 860 - 8
920 - 930 - 940 - 950 - 960 - 970 - 980 - 990 - 1

Reg Reps: 10 - 20 - 30 - 40 - 50 - 60 - 70 - 80 - 9
140 - 150 - 160 - 170 - 180 - 190 - 200 - 210 - 22
270 - 280 - 290 - 300 - 310 - 320 - 330 - 340 - 3
400 - 410 - 420 - 430 - 440 - 450 - 460 - 470 - 4
530 - 540 - 550 - 560 - 570 - 580 - 590 - 600 - 61
660 - 670 - 680 - 690 - 700 - 710 - 720 - 730 - 7
790 - 800 - 810 - 820 - 830 - 840 - 850 - 860 - 87
920 - 930 - 940 - 950 - 960 - 970 - 980 - 990 - 1
Notes: ________________________________

Reg Reps: 10 - 20 - 30 - 40 - 50 - 60 - 70 - 80 - 9
140 - 150 - 160 - 170 - 180 - 190 - 200 - 210 - 22
270 - 280 - 290 - 300 - 310 - 320 - 330 - 340 - 35
400 - 410 - 420 - 430 - 440 - 450 - 460 - 470 - 48
530 - 540 - 550 - 560 - 570 - 580 - 590 - 600 - 61
660 - 670 - 680 - 690 - 700 - 710 - 720 - 730 - 74
790 - 800 - 810 - 820 - 830 - 840 - 850 - 860 - 87
920 - 930 - 940 - 950 - 960 - 970 - 980 - 990 - 10
Notes: ________________________________

Reg Re
140 -
270 -
400 -
530 -
660 -
790 -
920 -
Notes:

Reg Re
140 -
270 -
400 -
530 -
660 -
790 -
920 -
Notes:

Reg Re
140 -
270 -
400 -
530 -
660 -
790 -
920 -
Notes:

MALFUNCTION DRILL #3: DOUBLE FEED

Holster Position: ______________ **Hand:** ______ **Start Date:** ________
Spare Magazine Position: ____________ **Firearm:** ___________________
Cover Garment: __________________________________ **Caliber:** ______
Notes: ____

____

____

Slow Reps: 10 - 20 - 30 - 40 - 50 - 60 - 70 - 80 - 90 - 100 - 110 - 120 - 130
140 - 150 - 160 - 170 - 180 - 190 - 200 - 210 - 220 - 230 - 240 - 250 - 260
270 - 280 - 290 - 300 - 310 - 320 - 330 - 340 - 350 - 360 - 370 - 380 - 390
400 - 410 - 420 - 430 - 440 - 450 - 460 - 470 - 480 - 490 - 500 - 510 - 520
530 - 540 - 550 - 560 - 570 - 580 - 590 - 600 - 610 - 620 - 630 - 640 - 650
660 - 670 - 680 - 690 - 700 - 710 - 720 - 730 - 740 - 750 - 760 - 770 - 780
790 - 800 - 810 - 820 - 830 - 840 - 850 - 860 - 870 - 880 - 890 - 900 - 910
920 - 930 - 940 - 950 - 960 - 970 - 980 - 990 - 1000

Reg Reps: 10 - 20 - 30 - 40 - 50 - 60 - 70 - 80 - 90 - 100 - 110 - 120 - 130
140 - 150 - 160 - 170 - 180 - 190 - 200 - 210 - 220 - 230 - 240 - 250 - 260
270 - 280 - 290 - 300 - 310 - 320 - 330 - 340 - 350 - 360 - 370 - 380 - 390
400 - 410 - 420 - 430 - 440 - 450 - 460 - 470 - 480 - 490 - 500 - 510 - 520
530 - 540 - 550 - 560 - 570 - 580 - 590 - 600 - 610 - 620 - 630 - 640 - 650
660 - 670 - 680 - 690 - 700 - 710 - 720 - 730 - 740 - 750 - 760 - 770 - 780
790 - 800 - 810 - 820 - 830 - 840 - 850 - 860 - 870 - 880 - 890 - 900 - 910
920 - 930 - 940 - 950 - 960 - 970 - 980 - 990 - 1000
Notes: ____

____

____

Reg Reps: 10 - 20 - 30 - 40 - 50 - 60 - 70 - 80 - 90 - 100 - 110 - 120 - 130
140 - 150 - 160 - 170 - 180 - 190 - 200 - 210 - 220 - 230 - 240 - 250 - 260
270 - 280 - 290 - 300 - 310 - 320 - 330 - 340 - 350 - 360 - 370 - 380 - 390
400 - 410 - 420 - 430 - 440 - 450 - 460 - 470 - 480 - 490 - 500 - 510 - 520
530 - 540 - 550 - 560 - 570 - 580 - 590 - 600 - 610 - 620 - 630 - 640 - 650
660 - 670 - 680 - 690 - 700 - 710 - 720 - 730 - 740 - 750 - 760 - 770 - 780
790 - 800 - 810 - 820 - 830 - 840 - 850 - 860 - 870 - 880 - 890 - 900 - 910
920 - 930 - 940 - 950 - 960 - 970 - 980 - 990 - 1000
Notes: ____

____

____

Reg Reps: 10 - 20 - 30 - 40 - 50 - 60 - 70 - 80 - 90 - 100 - 110 - 120 - 130
140 - 150 - 160 - 170 - 180 - 190 - 200 - 210 - 220 - 230 - 240 - 250 - 260
270 - 280 - 290 - 300 - 310 - 320 - 330 - 340 - 350 - 360 - 370 - 380 - 390
400 - 410 - 420 - 430 - 440 - 450 - 460 - 470 - 480 - 490 - 500 - 510 - 520
530 - 540 - 550 - 560 - 570 - 580 - 590 - 600 - 610 - 620 - 630 - 640 - 650
660 - 670 - 680 - 690 - 700 - 710 - 720 - 730 - 740 - 750 - 760 - 770 - 780
790 - 800 - 810 - 820 - 830 - 840 - 850 - 860 - 870 - 880 - 890 - 900 - 910
920 - 930 - 940 - 950 - 960 - 970 - 980 - 990 - 1000

Notes: ____

____

____

____

Reg Reps: 10 - 20 - 30 - 40 - 50 - 60 - 70 - 80 - 90 - 100 - 110 - 120 - 130
140 - 150 - 160 - 170 - 180 - 190 - 200 - 210 - 220 - 230 - 240 - 250 - 260
270 - 280 - 290 - 300 - 310 - 320 - 330 - 340 - 350 - 360 - 370 - 380 - 390
400 - 410 - 420 - 430 - 440 - 450 - 460 - 470 - 480 - 490 - 500 - 510 - 520
530 - 540 - 550 - 560 - 570 - 580 - 590 - 600 - 610 - 620 - 630 - 640 - 650
660 - 670 - 680 - 690 - 700 - 710 - 720 - 730 - 740 - 750 - 760 - 770 - 780
790 - 800 - 810 - 820 - 830 - 840 - 850 - 860 - 870 - 880 - 890 - 900 - 910
920 - 930 - 940 - 950 - 960 - 970 - 980 - 990 - 1000

Notes: ____

____

____

____

Reg Reps: 10 - 20 - 30 - 40 - 50 - 60 - 70 - 80 - 90 - 100 - 110 - 120 - 130
140 - 150 - 160 - 170 - 180 - 190 - 200 - 210 - 220 - 230 - 240 - 250 - 260
270 - 280 - 290 - 300 - 310 - 320 - 330 - 340 - 350 - 360 - 370 - 380 - 390
400 - 410 - 420 - 430 - 440 - 450 - 460 - 470 - 480 - 490 - 500 - 510 - 520
530 - 540 - 550 - 560 - 570 - 580 - 590 - 600 - 610 - 620 - 630 - 640 - 650
660 - 670 - 680 - 690 - 700 - 710 - 720 - 730 - 740 - 750 - 760 - 770 - 780
790 - 800 - 810 - 820 - 830 - 840 - 850 - 860 - 870 - 880 - 890 - 900 - 910
920 - 930 - 940 - 950 - 960 - 970 - 980 - 990 - 1000

Notes: ____

____

____

______________________________ **End Date:** ____________

LOADING THE MAGAZINE

Holster Position:_____________ **Hand:** ______ **Start Date:** ________
Spare Magazine Position: ___________ **Firearm:** __________________
Cover Garment: _______________________________ **Caliber:** ______
Notes:____

Slow Reps: 10 - 20 - 30 - 40 - 50 - 60 - 70 - 80 - 90 - 100 - 110 - 120 - 130
140 - 150 - 160 - 170 - 180 - 190 - 200 - 210 - 220 - 230 - 240 - 250 - 260
270 - 280 - 290 - 300 - 310 - 320 - 330 - 340 - 350 - 360 - 370 - 380 - 390
400 - 410 - 420 - 430 - 440 - 450 - 460 - 470 - 480 - 490 - 500 - 510 - 520
530 - 540 - 550 - 560 - 570 - 580 - 590 - 600 - 610 - 620 - 630 - 640 - 650
660 - 670 - 680 - 690 - 700 - 710 - 720 - 730 - 740 - 750 - 760 - 770 - 780
790 - 800 - 810 - 820 - 830 - 840 - 850 - 860 - 870 - 880 - 890 - 900 - 910
920 - 930 - 940 - 950 - 960 - 970 - 980 - 990 - 1000

Reg Reps: 10 - 20 - 30 - 40 - 50 - 60 - 70 - 80 - 90 - 100 - 110 - 120 - 130
140 - 150 - 160 - 170 - 180 - 190 - 200 - 210 - 220 - 230 - 240 - 250 - 260
270 - 280 - 290 - 300 - 310 - 320 - 330 - 340 - 350 - 360 - 370 - 380 - 390
400 - 410 - 420 - 430 - 440 - 450 - 460 - 470 - 480 - 490 - 500 - 510 - 520
530 - 540 - 550 - 560 - 570 - 580 - 590 - 600 - 610 - 620 - 630 - 640 - 650
660 - 670 - 680 - 690 - 700 - 710 - 720 - 730 - 740 - 750 - 760 - 770 - 780
790 - 800 - 810 - 820 - 830 - 840 - 850 - 860 - 870 - 880 - 890 - 900 - 910
920 - 930 - 940 - 950 - 960 - 970 - 980 - 990 - 1000
Notes: ____

Reg Reps: 10 - 20 - 30 - 40 - 50 - 60 - 70 - 80 - 90 - 100 - 110 - 120 - 130
140 - 150 - 160 - 170 - 180 - 190 - 200 - 210 - 220 - 230 - 240 - 250 - 260
270 - 280 - 290 - 300 - 310 - 320 - 330 - 340 - 350 - 360 - 370 - 380 - 390
400 - 410 - 420 - 430 - 440 - 450 - 460 - 470 - 480 - 490 - 500 - 510 - 520
530 - 540 - 550 - 560 - 570 - 580 - 590 - 600 - 610 - 620 - 630 - 640 - 650
660 - 670 - 680 - 690 - 700 - 710 - 720 - 730 - 740 - 750 - 760 - 770 - 780
790 - 800 - 810 - 820 - 830 - 840 - 850 - 860 - 870 - 880 - 890 - 900 - 910
920 - 930 - 940 - 950 - 960 - 970 - 980 - 990 - 1000
Notes: ____

Reg Reps: 10 - 20 - 30 - 40 - 50 - 60 - 70 - 80 - 90 - 100 - 110 - 120 - 130
140 - 150 - 160 - 170 - 180 - 190 - 200 - 210 - 220 - 230 - 240 - 250 - 260
270 - 280 - 290 - 300 - 310 - 320 - 330 - 340 - 350 - 360 - 370 - 380 - 390
400 - 410 - 420 - 430 - 440 - 450 - 460 - 470 - 480 - 490 - 500 - 510 - 520
530 - 540 - 550 - 560 - 570 - 580 - 590 - 600 - 610 - 620 - 630 - 640 - 650
660 - 670 - 680 - 690 - 700 - 710 - 720 - 730 - 740 - 750 - 760 - 770 - 780
790 - 800 - 810 - 820 - 830 - 840 - 850 - 860 - 870 - 880 - 890 - 900 - 910
920 - 930 - 940 - 950 - 960 - 970 - 980 - 990 - 1000

Notes: ____

____

____

____

Reg Reps: 10 - 20 - 30 - 40 - 50 - 60 - 70 - 80 - 90 - 100 - 110 - 120 - 130
140 - 150 - 160 - 170 - 180 - 190 - 200 - 210 - 220 - 230 - 240 - 250 - 260
270 - 280 - 290 - 300 - 310 - 320 - 330 - 340 - 350 - 360 - 370 - 380 - 390
400 - 410 - 420 - 430 - 440 - 450 - 460 - 470 - 480 - 490 - 500 - 510 - 520
530 - 540 - 550 - 560 - 570 - 580 - 590 - 600 - 610 - 620 - 630 - 640 - 650
660 - 670 - 680 - 690 - 700 - 710 - 720 - 730 - 740 - 750 - 760 - 770 - 780
790 - 800 - 810 - 820 - 830 - 840 - 850 - 860 - 870 - 880 - 890 - 900 - 910
920 - 930 - 940 - 950 - 960 - 970 - 980 - 990 - 1000

Notes: ____

____

____

____

Reg Reps: 10 - 20 - 30 - 40 - 50 - 60 - 70 - 80 - 90 - 100 - 110 - 120 - 130
140 - 150 - 160 - 170 - 180 - 190 - 200 - 210 - 220 - 230 - 240 - 250 - 260
270 - 280 - 290 - 300 - 310 - 320 - 330 - 340 - 350 - 360 - 370 - 380 - 390
400 - 410 - 420 - 430 - 440 - 450 - 460 - 470 - 480 - 490 - 500 - 510 - 520
530 - 540 - 550 - 560 - 570 - 580 - 590 - 600 - 610 - 620 - 630 - 640 - 650
660 - 670 - 680 - 690 - 700 - 710 - 720 - 730 - 740 - 750 - 760 - 770 - 780
790 - 800 - 810 - 820 - 830 - 840 - 850 - 860 - 870 - 880 - 890 - 900 - 910
920 - 930 - 940 - 950 - 960 - 970 - 980 - 990 - 1000

Notes: ____

____

____

______________________________ **End Date:** __________

0 - 20 - 30 - 40 - 50 - 60 - 70 - 80 - 90 - 100 - 110 - 120 - 130
160 - 170 - 180 - 190 - 200 - 210 - 220 - 230 - 240 - 250 - 260
290 - 300 - 310 - 320 - 330 - 340 - 350 - 360 - 370 - 380 - 390
420 - 430 - 440 - 450 - 460 - 470 - 480 - 490 - 500 - 510 - 520
550 - 560 - 570 - 580 - 590 - 600 - 610 - 620 - 630 - 640 - 650
680 - 690 - 700 - 710 - 720 - 730 - 740 - 750 - 760 - 770 - 780
810 - 820 - 830 - 840 - 850 - 860 - 870 - 880 - 890 - 900 - 910
940 - 950 - 960 - 970 - 980 - 990 - 1000

0 - 20 - 30 - 40 - 50 - 60 - 70 - 80 - 90 - 100 - 110 - 120 - 130
160 - 170 - 180 - 190 - 200 - 210 - 220 - 230 - 240 - 250 - 260
290 - 300 - 310 - 320 - 330 - 340 - 350 - 360 - 370 - 380 - 390
420 - 430 - 440 - 450 - 460 - 470 - 480 - 490 - 500 - 510 - 520
550 - 560 - 570 - 580 - 590 - 600 - 610 - 620 - 630 - 640 - 650
680 - 690 - 700 - 710 - 720 - 730 - 740 - 750 - 760 - 770 - 780
810 - 820 - 830 - 840 - 850 - 860 - 870 - 880 - 890 - 900 - 910
940 - 950 - 960 - 970 - 980 - 990 - 1000

10 - 20 - 30 - 40 - 50 - 60 - 70 - 80 - 90 - 100 - 110 - 120 - 130
160 - 170 - 180 - 190 - 200 - 210 - 220 - 230 - 240 - 250 - 260
290 - 300 - 310 - 320 - 330 - 340 - 350 - 360 - 370 - 380 - 390
420 - 430 - 440 - 450 - 460 - 470 - 480 - 490 - 500 - 510 - 520
550 - 560 - 570 - 580 - 590 - 600 - 610 - 620 - 630 - 640 - 650
680 - 690 - 700 - 710 - 720 - 730 - 740 - 750 - 760 - 770 - 780
810 - 820 - 830 - 840 - 850 - 860 - 870 - 880 - 890 - 900 - 910
940 - 950 - 960 - 970 - 980 - 990 - 1000

End Date: ____________

MAGAZINE CHAN

Holster Position:____________ **Han**
Spare Magazine Position: __________
Cover Garment: ____________
Notes:____________________

Slow Reps: 10 - 20 - 30 - 40 - 50 - 60 -
140 - 150 - 160 - 170 - 180 - 190 - 200
270 - 280 - 290 - 300 - 310 - 320 - 330
400 - 410 - 420 - 430 - 440 - 450 - 460
530 - 540 - 550 - 560 - 570 - 580 - 590
660 - 670 - 680 - 690 - 700 - 710 - 720
790 - 800 - 810 - 820 - 830 - 840 - 850
920 - 930 - 940 - 950 - 960 - 970 - 980

Reg Reps: 10 - 20 - 30 - 40 - 50 - 60 - 7
140 - 150 - 160 - 170 - 180 - 190 - 200 -
270 - 280 - 290 - 300 - 310 - 320 - 330 -
400 - 410 - 420 - 430 - 440 - 450 - 460 -
530 - 540 - 550 - 560 - 570 - 580 - 590 -
660 - 670 - 680 - 690 - 700 - 710 - 720 -
790 - 800 - 810 - 820 - 830 - 840 - 850 -
920 - 930 - 940 - 950 - 960 - 970 - 980 -
Notes: ____________________

Reg Reps: 10 - 20 - 30 - 40 - 50 - 60 - 70
140 - 150 - 160 - 170 - 180 - 190 - 200 -
270 - 280 - 290 - 300 - 310 - 320 - 330 -
400 - 410 - 420 - 430 - 440 - 450 - 460 -
530 - 540 - 550 - 560 - 570 - 580 - 590 -
660 - 670 - 680 - 690 - 700 - 710 - 720 -
790 - 800 - 810 - 820 - 830 - 840 - 850 -
920 - 930 - 940 - 950 - 960 - 970 - 980 -
Notes: ____________________

Reg Reps:
140 - 150 -
270 - 280 -
400 - 410 -
530 - 540 -
660 - 670 -
790 - 800 -
920 - 930 -
Notes: ____

Reg Reps:
140 - 150 -
270 - 280 -
400 - 410 -
530 - 540 -
660 - 670 -
790 - 800 -
920 - 930 -
Notes: ____

Reg Reps:
140 - 150
270 - 280
400 - 410
530 - 540
660 - 670
790 - 800
920 - 930
Notes: ____

MAGAZINE CHANGE: DOMINANT HAND ONLY

Holster Position:_____________ **Hand:** ______ **Start Date:** _________
Spare Magazine Position: ___________ **Firearm:** _________________
Cover Garment: ________________________________ **Caliber:** ______
Notes:____

____

____

Slow Reps: 10 - 20 - 30 - 40 - 50 - 60 - 70 - 80 - 90 - 100 - 110 - 120 - 130
140 - 150 - 160 - 170 - 180 - 190 - 200 - 210 - 220 - 230 - 240 - 250 - 260
270 - 280 - 290 - 300 - 310 - 320 - 330 - 340 - 350 - 360 - 370 - 380 - 390
400 - 410 - 420 - 430 - 440 - 450 - 460 - 470 - 480 - 490 - 500 - 510 - 520
530 - 540 - 550 - 560 - 570 - 580 - 590 - 600 - 610 - 620 - 630 - 640 - 650
660 - 670 - 680 - 690 - 700 - 710 - 720 - 730 - 740 - 750 - 760 - 770 - 780
790 - 800 - 810 - 820 - 830 - 840 - 850 - 860 - 870 - 880 - 890 - 900 - 910
920 - 930 - 940 - 950 - 960 - 970 - 980 - 990 - 1000

Reg Reps: 10 - 20 - 30 - 40 - 50 - 60 - 70 - 80 - 90 - 100 - 110 - 120 - 130
140 - 150 - 160 - 170 - 180 - 190 - 200 - 210 - 220 - 230 - 240 - 250 - 260
270 - 280 - 290 - 300 - 310 - 320 - 330 - 340 - 350 - 360 - 370 - 380 - 390
400 - 410 - 420 - 430 - 440 - 450 - 460 - 470 - 480 - 490 - 500 - 510 - 520
530 - 540 - 550 - 560 - 570 - 580 - 590 - 600 - 610 - 620 - 630 - 640 - 650
660 - 670 - 680 - 690 - 700 - 710 - 720 - 730 - 740 - 750 - 760 - 770 - 780
790 - 800 - 810 - 820 - 830 - 840 - 850 - 860 - 870 - 880 - 890 - 900 - 910
920 - 930 - 940 - 950 - 960 - 970 - 980 - 990 - 1000
Notes: ____

____

____

Reg Reps: 10 - 20 - 30 - 40 - 50 - 60 - 70 - 80 - 90 - 100 - 110 - 120 - 130
140 - 150 - 160 - 170 - 180 - 190 - 200 - 210 - 220 - 230 - 240 - 250 - 260
270 - 280 - 290 - 300 - 310 - 320 - 330 - 340 - 350 - 360 - 370 - 380 - 390
400 - 410 - 420 - 430 - 440 - 450 - 460 - 470 - 480 - 490 - 500 - 510 - 520
530 - 540 - 550 - 560 - 570 - 580 - 590 - 600 - 610 - 620 - 630 - 640 - 650
660 - 670 - 680 - 690 - 700 - 710 - 720 - 730 - 740 - 750 - 760 - 770 - 780
790 - 800 - 810 - 820 - 830 - 840 - 850 - 860 - 870 - 880 - 890 - 900 - 910
920 - 930 - 940 - 950 - 960 - 970 - 980 - 990 - 1000
Notes: ____

____

____

Reg Reps: 10 - 20 - 30 - 40 - 50 - 60 - 70 - 80 - 90 - 100 - 110 - 120 - 130 140 - 150 - 160 - 170 - 180 - 190 - 200 - 210 - 220 - 230 - 240 - 250 - 260 270 - 280 - 290 - 300 - 310 - 320 - 330 - 340 - 350 - 360 - 370 - 380 - 390 400 - 410 - 420 - 430 - 440 - 450 - 460 - 470 - 480 - 490 - 500 - 510 - 520 530 - 540 - 550 - 560 - 570 - 580 - 590 - 600 - 610 - 620 - 630 - 640 - 650 660 - 670 - 680 - 690 - 700 - 710 - 720 - 730 - 740 - 750 - 760 - 770 - 780 790 - 800 - 810 - 820 - 830 - 840 - 850 - 860 - 870 - 880 - 890 - 900 - 910 920 - 930 - 940 - 950 - 960 - 970 - 980 - 990 - 1000

Notes: ____

____

____

____

Reg Reps: 10 - 20 - 30 - 40 - 50 - 60 - 70 - 80 - 90 - 100 - 110 - 120 - 130 140 - 150 - 160 - 170 - 180 - 190 - 200 - 210 - 220 - 230 - 240 - 250 - 260 270 - 280 - 290 - 300 - 310 - 320 - 330 - 340 - 350 - 360 - 370 - 380 - 390 400 - 410 - 420 - 430 - 440 - 450 - 460 - 470 - 480 - 490 - 500 - 510 - 520 530 - 540 - 550 - 560 - 570 - 580 - 590 - 600 - 610 - 620 - 630 - 640 - 650 660 - 670 - 680 - 690 - 700 - 710 - 720 - 730 - 740 - 750 - 760 - 770 - 780 790 - 800 - 810 - 820 - 830 - 840 - 850 - 860 - 870 - 880 - 890 - 900 - 910 920 - 930 - 940 - 950 - 960 - 970 - 980 - 990 - 1000

Notes: ____

____

____

____

Reg Reps: 10 - 20 - 30 - 40 - 50 - 60 - 70 - 80 - 90 - 100 - 110 - 120 - 130 140 - 150 - 160 - 170 - 180 - 190 - 200 - 210 - 220 - 230 - 240 - 250 - 260 270 - 280 - 290 - 300 - 310 - 320 - 330 - 340 - 350 - 360 - 370 - 380 - 390 400 - 410 - 420 - 430 - 440 - 450 - 460 - 470 - 480 - 490 - 500 - 510 - 520 530 - 540 - 550 - 560 - 570 - 580 - 590 - 600 - 610 - 620 - 630 - 640 - 650 660 - 670 - 680 - 690 - 700 - 710 - 720 - 730 - 740 - 750 - 760 - 770 - 780 790 - 800 - 810 - 820 - 830 - 840 - 850 - 860 - 870 - 880 - 890 - 900 - 910 920 - 930 - 940 - 950 - 960 - 970 - 980 - 990 - 1000

Notes: ____

____

____

______________________________ **End Date:** __________

SLIDE RELEASE: MANUAL

Holster Position:______________ **Hand:** ______ **Start Date:** __________
Spare Magazine Position: ____________ **Firearm:** ____________________
Cover Garment: ____________________________________ **Caliber:** ______
Notes:____

Slow Reps: 10 - 20 - 30 - 40 - 50 - 60 - 70 - 80 - 90 - 100 - 110 - 120 - 130
140 - 150 - 160 - 170 - 180 - 190 - 200 - 210 - 220 - 230 - 240 - 250 - 260
270 - 280 - 290 - 300 - 310 - 320 - 330 - 340 - 350 - 360 - 370 - 380 - 390
400 - 410 - 420 - 430 - 440 - 450 - 460 - 470 - 480 - 490 - 500 - 510 - 520
530 - 540 - 550 - 560 - 570 - 580 - 590 - 600 - 610 - 620 - 630 - 640 - 650
660 - 670 - 680 - 690 - 700 - 710 - 720 - 730 - 740 - 750 - 760 - 770 - 780
790 - 800 - 810 - 820 - 830 - 840 - 850 - 860 - 870 - 880 - 890 - 900 - 910
920 - 930 - 940 - 950 - 960 - 970 - 980 - 990 - 1000

Reg Reps: 10 - 20 - 30 - 40 - 50 - 60 - 70 - 80 - 90 - 100 - 110 - 120 - 130
140 - 150 - 160 - 170 - 180 - 190 - 200 - 210 - 220 - 230 - 240 - 250 - 260
270 - 280 - 290 - 300 - 310 - 320 - 330 - 340 - 350 - 360 - 370 - 380 - 390
400 - 410 - 420 - 430 - 440 - 450 - 460 - 470 - 480 - 490 - 500 - 510 - 520
530 - 540 - 550 - 560 - 570 - 580 - 590 - 600 - 610 - 620 - 630 - 640 - 650
660 - 670 - 680 - 690 - 700 - 710 - 720 - 730 - 740 - 750 - 760 - 770 - 780
790 - 800 - 810 - 820 - 830 - 840 - 850 - 860 - 870 - 880 - 890 - 900 - 910
920 - 930 - 940 - 950 - 960 - 970 - 980 - 990 - 1000
Notes: ____

Reg Reps: 10 - 20 - 30 - 40 - 50 - 60 - 70 - 80 - 90 - 100 - 110 - 120 - 130
140 - 150 - 160 - 170 - 180 - 190 - 200 - 210 - 220 - 230 - 240 - 250 - 260
270 - 280 - 290 - 300 - 310 - 320 - 330 - 340 - 350 - 360 - 370 - 380 - 390
400 - 410 - 420 - 430 - 440 - 450 - 460 - 470 - 480 - 490 - 500 - 510 - 520
530 - 540 - 550 - 560 - 570 - 580 - 590 - 600 - 610 - 620 - 630 - 640 - 650
660 - 670 - 680 - 690 - 700 - 710 - 720 - 730 - 740 - 750 - 760 - 770 - 780
790 - 800 - 810 - 820 - 830 - 840 - 850 - 860 - 870 - 880 - 890 - 900 - 910
920 - 930 - 940 - 950 - 960 - 970 - 980 - 990 - 1000
Notes: ____

Reg Reps: 10 - 20 - 30 - 40 - 50 - 60 - 70 - 80 - 90 - 100 - 110 - 120 - 130 140 - 150 - 160 - 170 - 180 - 190 - 200 - 210 - 220 - 230 - 240 - 250 - 260 270 - 280 - 290 - 300 - 310 - 320 - 330 - 340 - 350 - 360 - 370 - 380 - 390 400 - 410 - 420 - 430 - 440 - 450 - 460 - 470 - 480 - 490 - 500 - 510 - 520 530 - 540 - 550 - 560 - 570 - 580 - 590 - 600 - 610 - 620 - 630 - 640 - 650 660 - 670 - 680 - 690 - 700 - 710 - 720 - 730 - 740 - 750 - 760 - 770 - 780 790 - 800 - 810 - 820 - 830 - 840 - 850 - 860 - 870 - 880 - 890 - 900 - 910 920 - 930 - 940 - 950 - 960 - 970 - 980 - 990 - 1000

Notes: ____

____

____

____

Reg Reps: 10 - 20 - 30 - 40 - 50 - 60 - 70 - 80 - 90 - 100 - 110 - 120 - 130 140 - 150 - 160 - 170 - 180 - 190 - 200 - 210 - 220 - 230 - 240 - 250 - 260 270 - 280 - 290 - 300 - 310 - 320 - 330 - 340 - 350 - 360 - 370 - 380 - 390 400 - 410 - 420 - 430 - 440 - 450 - 460 - 470 - 480 - 490 - 500 - 510 - 520 530 - 540 - 550 - 560 - 570 - 580 - 590 - 600 - 610 - 620 - 630 - 640 - 650 660 - 670 - 680 - 690 - 700 - 710 - 720 - 730 - 740 - 750 - 760 - 770 - 780 790 - 800 - 810 - 820 - 830 - 840 - 850 - 860 - 870 - 880 - 890 - 900 - 910 920 - 930 - 940 - 950 - 960 - 970 - 980 - 990 - 1000

Notes: ____

____

____

____

Reg Reps: 10 - 20 - 30 - 40 - 50 - 60 - 70 - 80 - 90 - 100 - 110 - 120 - 130 140 - 150 - 160 - 170 - 180 - 190 - 200 - 210 - 220 - 230 - 240 - 250 - 260 270 - 280 - 290 - 300 - 310 - 320 - 330 - 340 - 350 - 360 - 370 - 380 - 390 400 - 410 - 420 - 430 - 440 - 450 - 460 - 470 - 480 - 490 - 500 - 510 - 520 530 - 540 - 550 - 560 - 570 - 580 - 590 - 600 - 610 - 620 - 630 - 640 - 650 660 - 670 - 680 - 690 - 700 - 710 - 720 - 730 - 740 - 750 - 760 - 770 - 780 790 - 800 - 810 - 820 - 830 - 840 - 850 - 860 - 870 - 880 - 890 - 900 - 910 920 - 930 - 940 - 950 - 960 - 970 - 980 - 990 - 1000

Notes: ____

____

____

____________________________________ **End Date:** ____________

SLIDE RELEASE: SLIDE STOP

Holster Position: ______________ **Hand:** ______ **Start Date:** _________
Spare Magazine Position: ___________ **Firearm:** _________________
Cover Garment: ________________________________ **Caliber:** ______
Notes: ____

____

____

Slow Reps: 10 - 20 - 30 - 40 - 50 - 60 - 70 - 80 - 90 - 100 - 110 - 120 - 130
140 - 150 - 160 - 170 - 180 - 190 - 200 - 210 - 220 - 230 - 240 - 250 - 260
270 - 280 - 290 - 300 - 310 - 320 - 330 - 340 - 350 - 360 - 370 - 380 - 390
400 - 410 - 420 - 430 - 440 - 450 - 460 - 470 - 480 - 490 - 500 - 510 - 520
530 - 540 - 550 - 560 - 570 - 580 - 590 - 600 - 610 - 620 - 630 - 640 - 650
660 - 670 - 680 - 690 - 700 - 710 - 720 - 730 - 740 - 750 - 760 - 770 - 780
790 - 800 - 810 - 820 - 830 - 840 - 850 - 860 - 870 - 880 - 890 - 900 - 910
920 - 930 - 940 - 950 - 960 - 970 - 980 - 990 - 1000

Reg Reps: 10 - 20 - 30 - 40 - 50 - 60 - 70 - 80 - 90 - 100 - 110 - 120 - 130
140 - 150 - 160 - 170 - 180 - 190 - 200 - 210 - 220 - 230 - 240 - 250 - 260
270 - 280 - 290 - 300 - 310 - 320 - 330 - 340 - 350 - 360 - 370 - 380 - 390
400 - 410 - 420 - 430 - 440 - 450 - 460 - 470 - 480 - 490 - 500 - 510 - 520
530 - 540 - 550 - 560 - 570 - 580 - 590 - 600 - 610 - 620 - 630 - 640 - 650
660 - 670 - 680 - 690 - 700 - 710 - 720 - 730 - 740 - 750 - 760 - 770 - 780
790 - 800 - 810 - 820 - 830 - 840 - 850 - 860 - 870 - 880 - 890 - 900 - 910
920 - 930 - 940 - 950 - 960 - 970 - 980 - 990 - 1000
Notes: ____

____

____

Reg Reps: 10 - 20 - 30 - 40 - 50 - 60 - 70 - 80 - 90 - 100 - 110 - 120 - 130
140 - 150 - 160 - 170 - 180 - 190 - 200 - 210 - 220 - 230 - 240 - 250 - 260
270 - 280 - 290 - 300 - 310 - 320 - 330 - 340 - 350 - 360 - 370 - 380 - 390
400 - 410 - 420 - 430 - 440 - 450 - 460 - 470 - 480 - 490 - 500 - 510 - 520
530 - 540 - 550 - 560 - 570 - 580 - 590 - 600 - 610 - 620 - 630 - 640 - 650
660 - 670 - 680 - 690 - 700 - 710 - 720 - 730 - 740 - 750 - 760 - 770 - 780
790 - 800 - 810 - 820 - 830 - 840 - 850 - 860 - 870 - 880 - 890 - 900 - 910
920 - 930 - 940 - 950 - 960 - 970 - 980 - 990 - 1000
Notes: ____

____

____

Reg Reps: 10 - 20 - 30 - 40 - 50 - 60 - 70 - 80 - 90 - 100 - 110 - 120 - 130
140 - 150 - 160 - 170 - 180 - 190 - 200 - 210 - 220 - 230 - 240 - 250 - 260
270 - 280 - 290 - 300 - 310 - 320 - 330 - 340 - 350 - 360 - 370 - 380 - 390
400 - 410 - 420 - 430 - 440 - 450 - 460 - 470 - 480 - 490 - 500 - 510 - 52
530 - 540 - 550 - 560 - 570 - 580 - 590 - 600 - 610 - 620 - 630 - 640 - 650
660 - 670 - 680 - 690 - 700 - 710 - 720 - 730 - 740 - 750 - 760 - 770 - 780
790 - 800 - 810 - 820 - 830 - 840 - 850 - 860 - 870 - 880 - 890 - 900 - 910
920 - 930 - 940 - 950 - 960 - 970 - 980 - 990 - 1000

Notes: ____________________

Reg Reps: 10 - 20 - 30 - 40 - 50 - 60 - 70 - 80 - 90 - 100 - 110 - 120 - 130
140 - 150 - 160 - 170 - 180 - 190 - 200 - 210 - 220 - 230 - 240 - 250 - 260
270 - 280 - 290 - 300 - 310 - 320 - 330 - 340 - 350 - 360 - 370 - 380 - 390
400 - 410 - 420 - 430 - 440 - 450 - 460 - 470 - 480 - 490 - 500 - 510 - 520
530 - 540 - 550 - 560 - 570 - 580 - 590 - 600 - 610 - 620 - 630 - 640 - 650
660 - 670 - 680 - 690 - 700 - 710 - 720 - 730 - 740 - 750 - 760 - 770 - 780
790 - 800 - 810 - 820 - 830 - 840 - 850 - 860 - 870 - 880 - 890 - 900 - 910
920 - 930 - 940 - 950 - 960 - 970 - 980 - 990 - 1000

Notes: ____________________

Reg Reps: 10 - 20 - 30 - 40 - 50 - 60 - 70 - 80 - 90 - 100 - 110 - 120 - 130
140 - 150 - 160 - 170 - 180 - 190 - 200 - 210 - 220 - 230 - 240 - 250 - 260
270 - 280 - 290 - 300 - 310 - 320 - 330 - 340 - 350 - 360 - 370 - 380 - 390
400 - 410 - 420 - 430 - 440 - 450 - 460 - 470 - 480 - 490 - 500 - 510 - 520
530 - 540 - 550 - 560 - 570 - 580 - 590 - 600 - 610 - 620 - 630 - 640 - 650
660 - 670 - 680 - 690 - 700 - 710 - 720 - 730 - 740 - 750 - 760 - 770 - 780
790 - 800 - 810 - 820 - 830 - 840 - 850 - 860 - 870 - 880 - 890 - 900 - 910
920 - 930 - 940 - 950 - 960 - 970 - 980 - 990 - 1000

Notes: ____________________

End Date: ____________

DRAW AND SHOOT FROM THE HIP

Holster Position:______________ **Hand:** ______ **Start Date:** _________
Spare Magazine Position: ____________ **Firearm:** __________________
Cover Garment: ___________________________________ **Caliber:** ______
Notes:____

____

____

Slow Reps: 10 - 20 - 30 - 40 - 50 - 60 - 70 - 80 - 90 - 100 - 110 - 120 - 130
140 - 150 - 160 - 170 - 180 - 190 - 200 - 210 - 220 - 230 - 240 - 250 - 260
270 - 280 - 290 - 300 - 310 - 320 - 330 - 340 - 350 - 360 - 370 - 380 - 390
400 - 410 - 420 - 430 - 440 - 450 - 460 - 470 - 480 - 490 - 500 - 510 - 520
530 - 540 - 550 - 560 - 570 - 580 - 590 - 600 - 610 - 620 - 630 - 640 - 650
660 - 670 - 680 - 690 - 700 - 710 - 720 - 730 - 740 - 750 - 760 - 770 - 780
790 - 800 - 810 - 820 - 830 - 840 - 850 - 860 - 870 - 880 - 890 - 900 - 910
920 - 930 - 940 - 950 - 960 - 970 - 980 - 990 - 1000

Reg Reps: 10 - 20 - 30 - 40 - 50 - 60 - 70 - 80 - 90 - 100 - 110 - 120 - 130
140 - 150 - 160 - 170 - 180 - 190 - 200 - 210 - 220 - 230 - 240 - 250 - 260
270 - 280 - 290 - 300 - 310 - 320 - 330 - 340 - 350 - 360 - 370 - 380 - 390
400 - 410 - 420 - 430 - 440 - 450 - 460 - 470 - 480 - 490 - 500 - 510 - 520
530 - 540 - 550 - 560 - 570 - 580 - 590 - 600 - 610 - 620 - 630 - 640 - 650
660 - 670 - 680 - 690 - 700 - 710 - 720 - 730 - 740 - 750 - 760 - 770 - 780
790 - 800 - 810 - 820 - 830 - 840 - 850 - 860 - 870 - 880 - 890 - 900 - 910
920 - 930 - 940 - 950 - 960 - 970 - 980 - 990 - 1000
Notes: ____

____

____

Reg Reps: 10 - 20 - 30 - 40 - 50 - 60 - 70 - 80 - 90 - 100 - 110 - 120 - 130
140 - 150 - 160 - 170 - 180 - 190 - 200 - 210 - 220 - 230 - 240 - 250 - 260
270 - 280 - 290 - 300 - 310 - 320 - 330 - 340 - 350 - 360 - 370 - 380 - 390
400 - 410 - 420 - 430 - 440 - 450 - 460 - 470 - 480 - 490 - 500 - 510 - 520
530 - 540 - 550 - 560 - 570 - 580 - 590 - 600 - 610 - 620 - 630 - 640 - 650
660 - 670 - 680 - 690 - 700 - 710 - 720 - 730 - 740 - 750 - 760 - 770 - 780
790 - 800 - 810 - 820 - 830 - 840 - 850 - 860 - 870 - 880 - 890 - 900 - 910
920 - 930 - 940 - 950 - 960 - 970 - 980 - 990 - 1000
Notes: ____

____

____

Reg Reps: 10 - 20 - 30 - 40 - 50 - 60 - 70 - 80 - 90 - 100 - 110 - 120 - 130
140 - 150 - 160 - 170 - 180 - 190 - 200 - 210 - 220 - 230 - 240 - 250 - 260
270 - 280 - 290 - 300 - 310 - 320 - 330 - 340 - 350 - 360 - 370 - 380 - 390
400 - 410 - 420 - 430 - 440 - 450 - 460 - 470 - 480 - 490 - 500 - 510 - 520
530 - 540 - 550 - 560 - 570 - 580 - 590 - 600 - 610 - 620 - 630 - 640 - 650
660 - 670 - 680 - 690 - 700 - 710 - 720 - 730 - 740 - 750 - 760 - 770 - 780
790 - 800 - 810 - 820 - 830 - 840 - 850 - 860 - 870 - 880 - 890 - 900 - 910
920 - 930 - 940 - 950 - 960 - 970 - 980 - 990 - 1000

Notes: ____

____

____

____

Reg Reps: 10 - 20 - 30 - 40 - 50 - 60 - 70 - 80 - 90 - 100 - 110 - 120 - 130
140 - 150 - 160 - 170 - 180 - 190 - 200 - 210 - 220 - 230 - 240 - 250 - 260
270 - 280 - 290 - 300 - 310 - 320 - 330 - 340 - 350 - 360 - 370 - 380 - 390
400 - 410 - 420 - 430 - 440 - 450 - 460 - 470 - 480 - 490 - 500 - 510 - 520
530 - 540 - 550 - 560 - 570 - 580 - 590 - 600 - 610 - 620 - 630 - 640 - 650
660 - 670 - 680 - 690 - 700 - 710 - 720 - 730 - 740 - 750 - 760 - 770 - 780
790 - 800 - 810 - 820 - 830 - 840 - 850 - 860 - 870 - 880 - 890 - 900 - 910
920 - 930 - 940 - 950 - 960 - 970 - 980 - 990 - 1000

Notes: ____

____

____

____

Reg Reps: 10 - 20 - 30 - 40 - 50 - 60 - 70 - 80 - 90 - 100 - 110 - 120 - 130
140 - 150 - 160 - 170 - 180 - 190 - 200 - 210 - 220 - 230 - 240 - 250 - 260
270 - 280 - 290 - 300 - 310 - 320 - 330 - 340 - 350 - 360 - 370 - 380 - 390
400 - 410 - 420 - 430 - 440 - 450 - 460 - 470 - 480 - 490 - 500 - 510 - 520
530 - 540 - 550 - 560 - 570 - 580 - 590 - 600 - 610 - 620 - 630 - 640 - 650
660 - 670 - 680 - 690 - 700 - 710 - 720 - 730 - 740 - 750 - 760 - 770 - 780
790 - 800 - 810 - 820 - 830 - 840 - 850 - 860 - 870 - 880 - 890 - 900 - 910
920 - 930 - 940 - 950 - 960 - 970 - 980 - 990 - 1000

Notes: ____

____

____

____ **End Date:** ____________

DRAW WHILE DROPPING TO ONE KNEE

Holster Position:______________ **Hand:** ______ **Start Date:** _________
Spare Magazine Position: ___________ **Firearm:** ______________
Cover Garment: ________________________________ **Caliber:** ______
Notes:____

____

____

Slow Reps: 10 - 20 - 30 - 40 - 50 - 60 - 70 - 80 - 90 - 100 - 110 - 120 - 130
140 - 150 - 160 - 170 - 180 - 190 - 200 - 210 - 220 - 230 - 240 - 250 - 260
270 - 280 - 290 - 300 - 310 - 320 - 330 - 340 - 350 - 360 - 370 - 380 - 390
400 - 410 - 420 - 430 - 440 - 450 - 460 - 470 - 480 - 490 - 500 - 510 - 520
530 - 540 - 550 - 560 - 570 - 580 - 590 - 600 - 610 - 620 - 630 - 640 - 650
660 - 670 - 680 - 690 - 700 - 710 - 720 - 730 - 740 - 750 - 760 - 770 - 780
790 - 800 - 810 - 820 - 830 - 840 - 850 - 860 - 870 - 880 - 890 - 900 - 910
920 - 930 - 940 - 950 - 960 - 970 - 980 - 990 - 1000

Reg Reps: 10 - 20 - 30 - 40 - 50 - 60 - 70 - 80 - 90 - 100 - 110 - 120 - 130
140 - 150 - 160 - 170 - 180 - 190 - 200 - 210 - 220 - 230 - 240 - 250 - 260
270 - 280 - 290 - 300 - 310 - 320 - 330 - 340 - 350 - 360 - 370 - 380 - 390
400 - 410 - 420 - 430 - 440 - 450 - 460 - 470 - 480 - 490 - 500 - 510 - 520
530 - 540 - 550 - 560 - 570 - 580 - 590 - 600 - 610 - 620 - 630 - 640 - 650
660 - 670 - 680 - 690 - 700 - 710 - 720 - 730 - 740 - 750 - 760 - 770 - 780
790 - 800 - 810 - 820 - 830 - 840 - 850 - 860 - 870 - 880 - 890 - 900 - 910
920 - 930 - 940 - 950 - 960 - 970 - 980 - 990 - 1000
Notes: ____

____

____

Reg Reps: 10 - 20 - 30 - 40 - 50 - 60 - 70 - 80 - 90 - 100 - 110 - 120 - 130
140 - 150 - 160 - 170 - 180 - 190 - 200 - 210 - 220 - 230 - 240 - 250 - 260
270 - 280 - 290 - 300 - 310 - 320 - 330 - 340 - 350 - 360 - 370 - 380 - 390
400 - 410 - 420 - 430 - 440 - 450 - 460 - 470 - 480 - 490 - 500 - 510 - 520
530 - 540 - 550 - 560 - 570 - 580 - 590 - 600 - 610 - 620 - 630 - 640 - 650
660 - 670 - 680 - 690 - 700 - 710 - 720 - 730 - 740 - 750 - 760 - 770 - 780
790 - 800 - 810 - 820 - 830 - 840 - 850 - 860 - 870 - 880 - 890 - 900 - 910
920 - 930 - 940 - 950 - 960 - 970 - 980 - 990 - 1000
Notes: ____

____

____

Reg Reps: 10 - 20 - 30 - 40 - 50 - 60 - 70 - 80 - 90 - 100 - 110 - 120 - 130
140 - 150 - 160 - 170 - 180 - 190 - 200 - 210 - 220 - 230 - 240 - 250 - 260
270 - 280 - 290 - 300 - 310 - 320 - 330 - 340 - 350 - 360 - 370 - 380 - 390
400 - 410 - 420 - 430 - 440 - 450 - 460 - 470 - 480 - 490 - 500 - 510 - 520
530 - 540 - 550 - 560 - 570 - 580 - 590 - 600 - 610 - 620 - 630 - 640 - 650
660 - 670 - 680 - 690 - 700 - 710 - 720 - 730 - 740 - 750 - 760 - 770 - 780
790 - 800 - 810 - 820 - 830 - 840 - 850 - 860 - 870 - 880 - 890 - 900 - 910
920 - 930 - 940 - 950 - 960 - 970 - 980 - 990 - 1000

Notes: ____

____

____

____

Reg Reps: 10 - 20 - 30 - 40 - 50 - 60 - 70 - 80 - 90 - 100 - 110 - 120 - 130
140 - 150 - 160 - 170 - 180 - 190 - 200 - 210 - 220 - 230 - 240 - 250 - 260
270 - 280 - 290 - 300 - 310 - 320 - 330 - 340 - 350 - 360 - 370 - 380 - 390
400 - 410 - 420 - 430 - 440 - 450 - 460 - 470 - 480 - 490 - 500 - 510 - 520
530 - 540 - 550 - 560 - 570 - 580 - 590 - 600 - 610 - 620 - 630 - 640 - 650
660 - 670 - 680 - 690 - 700 - 710 - 720 - 730 - 740 - 750 - 760 - 770 - 780
790 - 800 - 810 - 820 - 830 - 840 - 850 - 860 - 870 - 880 - 890 - 900 - 910
920 - 930 - 940 - 950 - 960 - 970 - 980 - 990 - 1000

Notes: ____

____

____

____

Reg Reps: 10 - 20 - 30 - 40 - 50 - 60 - 70 - 80 - 90 - 100 - 110 - 120 - 130
140 - 150 - 160 - 170 - 180 - 190 - 200 - 210 - 220 - 230 - 240 - 250 - 260
270 - 280 - 290 - 300 - 310 - 320 - 330 - 340 - 350 - 360 - 370 - 380 - 390
400 - 410 - 420 - 430 - 440 - 450 - 460 - 470 - 480 - 490 - 500 - 510 - 520
530 - 540 - 550 - 560 - 570 - 580 - 590 - 600 - 610 - 620 - 630 - 640 - 650
660 - 670 - 680 - 690 - 700 - 710 - 720 - 730 - 740 - 750 - 760 - 770 - 780
790 - 800 - 810 - 820 - 830 - 840 - 850 - 860 - 870 - 880 - 890 - 900 - 910
920 - 930 - 940 - 950 - 960 - 970 - 980 - 990 - 1000

Notes: ____

____

____

______________________________________**End Date:**__________

DRAW WHILE MOVING: FORWARD

Holster Position:______________ **Hand:** ______ **Start Date:** _________
Spare Magazine Position: ____________ **Firearm:** ___________________
Cover Garment: ___________________________________ **Caliber:** ______
Notes:____
____
____

Slow Reps: 10 - 20 - 30 - 40 - 50 - 60 - 70 - 80 - 90 - 100 - 110 - 120 - 130
140 - 150 - 160 - 170 - 180 - 190 - 200 - 210 - 220 - 230 - 240 - 250 - 260
270 - 280 - 290 - 300 - 310 - 320 - 330 - 340 - 350 - 360 - 370 - 380 - 390
400 - 410 - 420 - 430 - 440 - 450 - 460 - 470 - 480 - 490 - 500 - 510 - 520
530 - 540 - 550 - 560 - 570 - 580 - 590 - 600 - 610 - 620 - 630 - 640 - 650
660 - 670 - 680 - 690 - 700 - 710 - 720 - 730 - 740 - 750 - 760 - 770 - 780
790 - 800 - 810 - 820 - 830 - 840 - 850 - 860 - 870 - 880 - 890 - 900 - 910
920 - 930 - 940 - 950 - 960 - 970 - 980 - 990 - 1000

Reg Reps: 10 - 20 - 30 - 40 - 50 - 60 - 70 - 80 - 90 - 100 - 110 - 120 - 130
140 - 150 - 160 - 170 - 180 - 190 - 200 - 210 - 220 - 230 - 240 - 250 - 260
270 - 280 - 290 - 300 - 310 - 320 - 330 - 340 - 350 - 360 - 370 - 380 - 390
400 - 410 - 420 - 430 - 440 - 450 - 460 - 470 - 480 - 490 - 500 - 510 - 520
530 - 540 - 550 - 560 - 570 - 580 - 590 - 600 - 610 - 620 - 630 - 640 - 650
660 - 670 - 680 - 690 - 700 - 710 - 720 - 730 - 740 - 750 - 760 - 770 - 780
790 - 800 - 810 - 820 - 830 - 840 - 850 - 860 - 870 - 880 - 890 - 900 - 910
920 - 930 - 940 - 950 - 960 - 970 - 980 - 990 - 1000
Notes: ____
____
____

Reg Reps: 10 - 20 - 30 - 40 - 50 - 60 - 70 - 80 - 90 - 100 - 110 - 120 - 130
140 - 150 - 160 - 170 - 180 - 190 - 200 - 210 - 220 - 230 - 240 - 250 - 260
270 - 280 - 290 - 300 - 310 - 320 - 330 - 340 - 350 - 360 - 370 - 380 - 390
400 - 410 - 420 - 430 - 440 - 450 - 460 - 470 - 480 - 490 - 500 - 510 - 520
530 - 540 - 550 - 560 - 570 - 580 - 590 - 600 - 610 - 620 - 630 - 640 - 650
660 - 670 - 680 - 690 - 700 - 710 - 720 - 730 - 740 - 750 - 760 - 770 - 780
790 - 800 - 810 - 820 - 830 - 840 - 850 - 860 - 870 - 880 - 890 - 900 - 910
920 - 930 - 940 - 950 - 960 - 970 - 980 - 990 - 1000
Notes: ____
____
____

Reg Reps: 10 - 20 - 30 - 40 - 50 - 60 - 70 - 80 - 90 - 100 - 110 - 120 - 130
140 - 150 - 160 - 170 - 180 - 190 - 200 - 210 - 220 - 230 - 240 - 250 - 260
270 - 280 - 290 - 300 - 310 - 320 - 330 - 340 - 350 - 360 - 370 - 380 - 390
400 - 410 - 420 - 430 - 440 - 450 - 460 - 470 - 480 - 490 - 500 - 510 - 520
530 - 540 - 550 - 560 - 570 - 580 - 590 - 600 - 610 - 620 - 630 - 640 - 650
660 - 670 - 680 - 690 - 700 - 710 - 720 - 730 - 740 - 750 - 760 - 770 - 780
790 - 800 - 810 - 820 - 830 - 840 - 850 - 860 - 870 - 880 - 890 - 900 - 910
920 - 930 - 940 - 950 - 960 - 970 - 980 - 990 - 1000

Notes: ____________________

____

____

____

Reg Reps: 10 - 20 - 30 - 40 - 50 - 60 - 70 - 80 - 90 - 100 - 110 - 120 - 130
140 - 150 - 160 - 170 - 180 - 190 - 200 - 210 - 220 - 230 - 240 - 250 - 260
270 - 280 - 290 - 300 - 310 - 320 - 330 - 340 - 350 - 360 - 370 - 380 - 390
400 - 410 - 420 - 430 - 440 - 450 - 460 - 470 - 480 - 490 - 500 - 510 - 520
530 - 540 - 550 - 560 - 570 - 580 - 590 - 600 - 610 - 620 - 630 - 640 - 650
660 - 670 - 680 - 690 - 700 - 710 - 720 - 730 - 740 - 750 - 760 - 770 - 780
790 - 800 - 810 - 820 - 830 - 840 - 850 - 860 - 870 - 880 - 890 - 900 - 910
920 - 930 - 940 - 950 - 960 - 970 - 980 - 990 - 1000

Notes: ____________________

____

____

____

Reg Reps: 10 - 20 - 30 - 40 - 50 - 60 - 70 - 80 - 90 - 100 - 110 - 120 - 130
140 - 150 - 160 - 170 - 180 - 190 - 200 - 210 - 220 - 230 - 240 - 250 - 260
270 - 280 - 290 - 300 - 310 - 320 - 330 - 340 - 350 - 360 - 370 - 380 - 390
400 - 410 - 420 - 430 - 440 - 450 - 460 - 470 - 480 - 490 - 500 - 510 - 520
530 - 540 - 550 - 560 - 570 - 580 - 590 - 600 - 610 - 620 - 630 - 640 - 650
660 - 670 - 680 - 690 - 700 - 710 - 720 - 730 - 740 - 750 - 760 - 770 - 780
790 - 800 - 810 - 820 - 830 - 840 - 850 - 860 - 870 - 880 - 890 - 900 - 910
920 - 930 - 940 - 950 - 960 - 970 - 980 - 990 - 1000

Notes: ____________________

____

____

____________________ **End Date:** __________

DRAW WHILE MOVING: BACKWARD

Holster Position:______________ **Hand:** ______ **Start Date:** _________
Spare Magazine Position: ___________ **Firearm:** ________________
Cover Garment: ________________________________ **Caliber:** ______
Notes:____

____

____

Slow Reps: 10 - 20 - 30 - 40 - 50 - 60 - 70 - 80 - 90 - 100 - 110 - 120 - 130
140 - 150 - 160 - 170 - 180 - 190 - 200 - 210 - 220 - 230 - 240 - 250 - 260
270 - 280 - 290 - 300 - 310 - 320 - 330 - 340 - 350 - 360 - 370 - 380 - 390
400 - 410 - 420 - 430 - 440 - 450 - 460 - 470 - 480 - 490 - 500 - 510 - 520
530 - 540 - 550 - 560 - 570 - 580 - 590 - 600 - 610 - 620 - 630 - 640 - 650
660 - 670 - 680 - 690 - 700 - 710 - 720 - 730 - 740 - 750 - 760 - 770 - 780
790 - 800 - 810 - 820 - 830 - 840 - 850 - 860 - 870 - 880 - 890 - 900 - 910
920 - 930 - 940 - 950 - 960 - 970 - 980 - 990 - 1000

Reg Reps: 10 - 20 - 30 - 40 - 50 - 60 - 70 - 80 - 90 - 100 - 110 - 120 - 130
140 - 150 - 160 - 170 - 180 - 190 - 200 - 210 - 220 - 230 - 240 - 250 - 260
270 - 280 - 290 - 300 - 310 - 320 - 330 - 340 - 350 - 360 - 370 - 380 - 390
400 - 410 - 420 - 430 - 440 - 450 - 460 - 470 - 480 - 490 - 500 - 510 - 520
530 - 540 - 550 - 560 - 570 - 580 - 590 - 600 - 610 - 620 - 630 - 640 - 650
660 - 670 - 680 - 690 - 700 - 710 - 720 - 730 - 740 - 750 - 760 - 770 - 780
790 - 800 - 810 - 820 - 830 - 840 - 850 - 860 - 870 - 880 - 890 - 900 - 910
920 - 930 - 940 - 950 - 960 - 970 - 980 - 990 - 1000
Notes: ____

____

____

Reg Reps: 10 - 20 - 30 - 40 - 50 - 60 - 70 - 80 - 90 - 100 - 110 - 120 - 130
140 - 150 - 160 - 170 - 180 - 190 - 200 - 210 - 220 - 230 - 240 - 250 - 260
270 - 280 - 290 - 300 - 310 - 320 - 330 - 340 - 350 - 360 - 370 - 380 - 390
400 - 410 - 420 - 430 - 440 - 450 - 460 - 470 - 480 - 490 - 500 - 510 - 520
530 - 540 - 550 - 560 - 570 - 580 - 590 - 600 - 610 - 620 - 630 - 640 - 650
660 - 670 - 680 - 690 - 700 - 710 - 720 - 730 - 740 - 750 - 760 - 770 - 780
790 - 800 - 810 - 820 - 830 - 840 - 850 - 860 - 870 - 880 - 890 - 900 - 910
920 - 930 - 940 - 950 - 960 - 970 - 980 - 990 - 1000
Notes: ____

____

____

Reg Reps: 10 - 20 - 30 - 40 - 50 - 60 - 70 - 80 - 90 - 100 - 110 - 120 - 130
140 - 150 - 160 - 170 - 180 - 190 - 200 - 210 - 220 - 230 - 240 - 250 - 260
270 - 280 - 290 - 300 - 310 - 320 - 330 - 340 - 350 - 360 - 370 - 380 - 390
400 - 410 - 420 - 430 - 440 - 450 - 460 - 470 - 480 - 490 - 500 - 510 - 520
530 - 540 - 550 - 560 - 570 - 580 - 590 - 600 - 610 - 620 - 630 - 640 - 650
660 - 670 - 680 - 690 - 700 - 710 - 720 - 730 - 740 - 750 - 760 - 770 - 780
790 - 800 - 810 - 820 - 830 - 840 - 850 - 860 - 870 - 880 - 890 - 900 - 910
920 - 930 - 940 - 950 - 960 - 970 - 980 - 990 - 1000

Notes: ____

____

____

____

Reg Reps: 10 - 20 - 30 - 40 - 50 - 60 - 70 - 80 - 90 - 100 - 110 - 120 - 130
140 - 150 - 160 - 170 - 180 - 190 - 200 - 210 - 220 - 230 - 240 - 250 - 260
270 - 280 - 290 - 300 - 310 - 320 - 330 - 340 - 350 - 360 - 370 - 380 - 390
400 - 410 - 420 - 430 - 440 - 450 - 460 - 470 - 480 - 490 - 500 - 510 - 520
530 - 540 - 550 - 560 - 570 - 580 - 590 - 600 - 610 - 620 - 630 - 640 - 650
660 - 670 - 680 - 690 - 700 - 710 - 720 - 730 - 740 - 750 - 760 - 770 - 780
790 - 800 - 810 - 820 - 830 - 840 - 850 - 860 - 870 - 880 - 890 - 900 - 910
920 - 930 - 940 - 950 - 960 - 970 - 980 - 990 - 1000

Notes: ____

____

____

____

Reg Reps: 10 - 20 - 30 - 40 - 50 - 60 - 70 - 80 - 90 - 100 - 110 - 120 - 130
140 - 150 - 160 - 170 - 180 - 190 - 200 - 210 - 220 - 230 - 240 - 250 - 260
270 - 280 - 290 - 300 - 310 - 320 - 330 - 340 - 350 - 360 - 370 - 380 - 390
400 - 410 - 420 - 430 - 440 - 450 - 460 - 470 - 480 - 490 - 500 - 510 - 520
530 - 540 - 550 - 560 - 570 - 580 - 590 - 600 - 610 - 620 - 630 - 640 - 650
660 - 670 - 680 - 690 - 700 - 710 - 720 - 730 - 740 - 750 - 760 - 770 - 780
790 - 800 - 810 - 820 - 830 - 840 - 850 - 860 - 870 - 880 - 890 - 900 - 910
920 - 930 - 940 - 950 - 960 - 970 - 980 - 990 - 1000

Notes: ____

____

____

______________________________ **End Date:** ____________

DRAW WHILE MOVING: TO THE LEFT

Holster Position:_____________ **Hand:** ______ **Start Date:** _________
Spare Magazine Position: ___________ **Firearm:** __________________
Cover Garment: _________________________________ **Caliber:** ______
Notes:____

____

____

Slow Reps: 10 - 20 - 30 - 40 - 50 - 60 - 70 - 80 - 90 - 100 - 110 - 120 - 130
140 - 150 - 160 - 170 - 180 - 190 - 200 - 210 - 220 - 230 - 240 - 250 - 260
270 - 280 - 290 - 300 - 310 - 320 - 330 - 340 - 350 - 360 - 370 - 380 - 390
400 - 410 - 420 - 430 - 440 - 450 - 460 - 470 - 480 - 490 - 500 - 510 - 520
530 - 540 - 550 - 560 - 570 - 580 - 590 - 600 - 610 - 620 - 630 - 640 - 650
660 - 670 - 680 - 690 - 700 - 710 - 720 - 730 - 740 - 750 - 760 - 770 - 780
790 - 800 - 810 - 820 - 830 - 840 - 850 - 860 - 870 - 880 - 890 - 900 - 910
920 - 930 - 940 - 950 - 960 - 970 - 980 - 990 - 1000

Reg Reps: 10 - 20 - 30 - 40 - 50 - 60 - 70 - 80 - 90 - 100 - 110 - 120 - 130
140 - 150 - 160 - 170 - 180 - 190 - 200 - 210 - 220 - 230 - 240 - 250 - 260
270 - 280 - 290 - 300 - 310 - 320 - 330 - 340 - 350 - 360 - 370 - 380 - 390
400 - 410 - 420 - 430 - 440 - 450 - 460 - 470 - 480 - 490 - 500 - 510 - 520
530 - 540 - 550 - 560 - 570 - 580 - 590 - 600 - 610 - 620 - 630 - 640 - 650
660 - 670 - 680 - 690 - 700 - 710 - 720 - 730 - 740 - 750 - 760 - 770 - 780
790 - 800 - 810 - 820 - 830 - 840 - 850 - 860 - 870 - 880 - 890 - 900 - 910
920 - 930 - 940 - 950 - 960 - 970 - 980 - 990 - 1000
Notes: ____

____

____

Reg Reps: 10 - 20 - 30 - 40 - 50 - 60 - 70 - 80 - 90 - 100 - 110 - 120 - 130
140 - 150 - 160 - 170 - 180 - 190 - 200 - 210 - 220 - 230 - 240 - 250 - 260
270 - 280 - 290 - 300 - 310 - 320 - 330 - 340 - 350 - 360 - 370 - 380 - 390
400 - 410 - 420 - 430 - 440 - 450 - 460 - 470 - 480 - 490 - 500 - 510 - 520
530 - 540 - 550 - 560 - 570 - 580 - 590 - 600 - 610 - 620 - 630 - 640 - 650
660 - 670 - 680 - 690 - 700 - 710 - 720 - 730 - 740 - 750 - 760 - 770 - 780
790 - 800 - 810 - 820 - 830 - 840 - 850 - 860 - 870 - 880 - 890 - 900 - 910
920 - 930 - 940 - 950 - 960 - 970 - 980 - 990 - 1000
Notes: ____

____

____

Reg Reps: 10 - 20 - 30 - 40 - 50 - 60 - 70 - 80 - 90 - 100 - 110 - 120 - 130
140 - 150 - 160 - 170 - 180 - 190 - 200 - 210 - 220 - 230 - 240 - 250 - 260
270 - 280 - 290 - 300 - 310 - 320 - 330 - 340 - 350 - 360 - 370 - 380 - 390
400 - 410 - 420 - 430 - 440 - 450 - 460 - 470 - 480 - 490 - 500 - 510 - 520
530 - 540 - 550 - 560 - 570 - 580 - 590 - 600 - 610 - 620 - 630 - 640 - 650
660 - 670 - 680 - 690 - 700 - 710 - 720 - 730 - 740 - 750 - 760 - 770 - 780
790 - 800 - 810 - 820 - 830 - 840 - 850 - 860 - 870 - 880 - 890 - 900 - 910
920 - 930 - 940 - 950 - 960 - 970 - 980 - 990 - 1000

Notes: ____

____

____

____

Reg Reps: 10 - 20 - 30 - 40 - 50 - 60 - 70 - 80 - 90 - 100 - 110 - 120 - 130
140 - 150 - 160 - 170 - 180 - 190 - 200 - 210 - 220 - 230 - 240 - 250 - 260
270 - 280 - 290 - 300 - 310 - 320 - 330 - 340 - 350 - 360 - 370 - 380 - 390
400 - 410 - 420 - 430 - 440 - 450 - 460 - 470 - 480 - 490 - 500 - 510 - 520
530 - 540 - 550 - 560 - 570 - 580 - 590 - 600 - 610 - 620 - 630 - 640 - 650
660 - 670 - 680 - 690 - 700 - 710 - 720 - 730 - 740 - 750 - 760 - 770 - 780
790 - 800 - 810 - 820 - 830 - 840 - 850 - 860 - 870 - 880 - 890 - 900 - 910
920 - 930 - 940 - 950 - 960 - 970 - 980 - 990 - 1000

Notes: ____

____

____

____

Reg Reps: 10 - 20 - 30 - 40 - 50 - 60 - 70 - 80 - 90 - 100 - 110 - 120 - 130
140 - 150 - 160 - 170 - 180 - 190 - 200 - 210 - 220 - 230 - 240 - 250 - 260
270 - 280 - 290 - 300 - 310 - 320 - 330 - 340 - 350 - 360 - 370 - 380 - 390
400 - 410 - 420 - 430 - 440 - 450 - 460 - 470 - 480 - 490 - 500 - 510 - 520
530 - 540 - 550 - 560 - 570 - 580 - 590 - 600 - 610 - 620 - 630 - 640 - 650
660 - 670 - 680 - 690 - 700 - 710 - 720 - 730 - 740 - 750 - 760 - 770 - 780
790 - 800 - 810 - 820 - 830 - 840 - 850 - 860 - 870 - 880 - 890 - 900 - 910
920 - 930 - 940 - 950 - 960 - 970 - 980 - 990 - 1000

Notes: ____

____

____

______________________________ **End Date:** ____________

DRAW WHILE MOVING: TO THE RIGHT

Holster Position:______________ **Hand:** ______ **Start Date:** _________
Spare Magazine Position: ___________ **Firearm:** _________________
Cover Garment: ________________________________ **Caliber:** ______
Notes:____

____

____

Slow Reps: 10 - 20 - 30 - 40 - 50 - 60 - 70 - 80 - 90 - 100 - 110 - 120 - 130
140 - 150 - 160 - 170 - 180 - 190 - 200 - 210 - 220 - 230 - 240 - 250 - 260
270 - 280 - 290 - 300 - 310 - 320 - 330 - 340 - 350 - 360 - 370 - 380 - 390
400 - 410 - 420 - 430 - 440 - 450 - 460 - 470 - 480 - 490 - 500 - 510 - 520
530 - 540 - 550 - 560 - 570 - 580 - 590 - 600 - 610 - 620 - 630 - 640 - 650
660 - 670 - 680 - 690 - 700 - 710 - 720 - 730 - 740 - 750 - 760 - 770 - 780
790 - 800 - 810 - 820 - 830 - 840 - 850 - 860 - 870 - 880 - 890 - 900 - 910
920 - 930 - 940 - 950 - 960 - 970 - 980 - 990 - 1000

Reg Reps: 10 - 20 - 30 - 40 - 50 - 60 - 70 - 80 - 90 - 100 - 110 - 120 - 130
140 - 150 - 160 - 170 - 180 - 190 - 200 - 210 - 220 - 230 - 240 - 250 - 260
270 - 280 - 290 - 300 - 310 - 320 - 330 - 340 - 350 - 360 - 370 - 380 - 390
400 - 410 - 420 - 430 - 440 - 450 - 460 - 470 - 480 - 490 - 500 - 510 - 520
530 - 540 - 550 - 560 - 570 - 580 - 590 - 600 - 610 - 620 - 630 - 640 - 650
660 - 670 - 680 - 690 - 700 - 710 - 720 - 730 - 740 - 750 - 760 - 770 - 780
790 - 800 - 810 - 820 - 830 - 840 - 850 - 860 - 870 - 880 - 890 - 900 - 910
920 - 930 - 940 - 950 - 960 - 970 - 980 - 990 - 1000
Notes: ____

____

____

Reg Reps: 10 - 20 - 30 - 40 - 50 - 60 - 70 - 80 - 90 - 100 - 110 - 120 - 130
140 - 150 - 160 - 170 - 180 - 190 - 200 - 210 - 220 - 230 - 240 - 250 - 260
270 - 280 - 290 - 300 - 310 - 320 - 330 - 340 - 350 - 360 - 370 - 380 - 390
400 - 410 - 420 - 430 - 440 - 450 - 460 - 470 - 480 - 490 - 500 - 510 - 520
530 - 540 - 550 - 560 - 570 - 580 - 590 - 600 - 610 - 620 - 630 - 640 - 650
660 - 670 - 680 - 690 - 700 - 710 - 720 - 730 - 740 - 750 - 760 - 770 - 780
790 - 800 - 810 - 820 - 830 - 840 - 850 - 860 - 870 - 880 - 890 - 900 - 910
920 - 930 - 940 - 950 - 960 - 970 - 980 - 990 - 1000
Notes: ____

____

____

Reg Reps: 10 - 20 - 30 - 40 - 50 - 60 - 70 - 80 - 90 - 100 - 110 - 120 - 130 - 140 - 150 - 160 - 170 - 180 - 190 - 200 - 210 - 220 - 230 - 240 - 250 - 260 - 270 - 280 - 290 - 300 - 310 - 320 - 330 - 340 - 350 - 360 - 370 - 380 - 390 - 400 - 410 - 420 - 430 - 440 - 450 - 460 - 470 - 480 - 490 - 500 - 510 - 520 - 530 - 540 - 550 - 560 - 570 - 580 - 590 - 600 - 610 - 620 - 630 - 640 - 650 - 660 - 670 - 680 - 690 - 700 - 710 - 720 - 730 - 740 - 750 - 760 - 770 - 780 - 790 - 800 - 810 - 820 - 830 - 840 - 850 - 860 - 870 - 880 - 890 - 900 - 910 - 920 - 930 - 940 - 950 - 960 - 970 - 980 - 990 - 1000

Notes: ____________________

____

____

____

Reg Reps: 10 - 20 - 30 - 40 - 50 - 60 - 70 - 80 - 90 - 100 - 110 - 120 - 130 - 140 - 150 - 160 - 170 - 180 - 190 - 200 - 210 - 220 - 230 - 240 - 250 - 260 - 270 - 280 - 290 - 300 - 310 - 320 - 330 - 340 - 350 - 360 - 370 - 380 - 390 - 400 - 410 - 420 - 430 - 440 - 450 - 460 - 470 - 480 - 490 - 500 - 510 - 520 - 530 - 540 - 550 - 560 - 570 - 580 - 590 - 600 - 610 - 620 - 630 - 640 - 650 - 660 - 670 - 680 - 690 - 700 - 710 - 720 - 730 - 740 - 750 - 760 - 770 - 780 - 790 - 800 - 810 - 820 - 830 - 840 - 850 - 860 - 870 - 880 - 890 - 900 - 910 - 920 - 930 - 940 - 950 - 960 - 970 - 980 - 990 - 1000

Notes: ____________________

____

____

____

Reg Reps: 10 - 20 - 30 - 40 - 50 - 60 - 70 - 80 - 90 - 100 - 110 - 120 - 130 - 140 - 150 - 160 - 170 - 180 - 190 - 200 - 210 - 220 - 230 - 240 - 250 - 260 - 270 - 280 - 290 - 300 - 310 - 320 - 330 - 340 - 350 - 360 - 370 - 380 - 390 - 400 - 410 - 420 - 430 - 440 - 450 - 460 - 470 - 480 - 490 - 500 - 510 - 520 - 530 - 540 - 550 - 560 - 570 - 580 - 590 - 600 - 610 - 620 - 630 - 640 - 650 - 660 - 670 - 680 - 690 - 700 - 710 - 720 - 730 - 740 - 750 - 760 - 770 - 780 - 790 - 800 - 810 - 820 - 830 - 840 - 850 - 860 - 870 - 880 - 890 - 900 - 910 - 920 - 930 - 940 - 950 - 960 - 970 - 980 - 990 - 1000

Notes: ____________________

____

____

____________________ **End Date:** ____________

DRAW WHILE LYING ON YOUR BACK

Holster Position:_____________ **Hand:** ______ **Start Date:** _________
Spare Magazine Position: ___________ **Firearm:** __________________
Cover Garment: ________________________________ **Caliber:** ______
Notes:____

____

____

Slow Reps: 10 - 20 - 30 - 40 - 50 - 60 - 70 - 80 - 90 - 100 - 110 - 120 - 130
140 - 150 - 160 - 170 - 180 - 190 - 200 - 210 - 220 - 230 - 240 - 250 - 260
270 - 280 - 290 - 300 - 310 - 320 - 330 - 340 - 350 - 360 - 370 - 380 - 390
400 - 410 - 420 - 430 - 440 - 450 - 460 - 470 - 480 - 490 - 500 - 510 - 520
530 - 540 - 550 - 560 - 570 - 580 - 590 - 600 - 610 - 620 - 630 - 640 - 650
660 - 670 - 680 - 690 - 700 - 710 - 720 - 730 - 740 - 750 - 760 - 770 - 780
790 - 800 - 810 - 820 - 830 - 840 - 850 - 860 - 870 - 880 - 890 - 900 - 910
920 - 930 - 940 - 950 - 960 - 970 - 980 - 990 - 1000

Reg Reps: 10 - 20 - 30 - 40 - 50 - 60 - 70 - 80 - 90 - 100 - 110 - 120 - 130
140 - 150 - 160 - 170 - 180 - 190 - 200 - 210 - 220 - 230 - 240 - 250 - 260
270 - 280 - 290 - 300 - 310 - 320 - 330 - 340 - 350 - 360 - 370 - 380 - 390
400 - 410 - 420 - 430 - 440 - 450 - 460 - 470 - 480 - 490 - 500 - 510 - 520
530 - 540 - 550 - 560 - 570 - 580 - 590 - 600 - 610 - 620 - 630 - 640 - 650
660 - 670 - 680 - 690 - 700 - 710 - 720 - 730 - 740 - 750 - 760 - 770 - 780
790 - 800 - 810 - 820 - 830 - 840 - 850 - 860 - 870 - 880 - 890 - 900 - 910
920 - 930 - 940 - 950 - 960 - 970 - 980 - 990 - 1000
Notes: ____

____

____

Reg Reps: 10 - 20 - 30 - 40 - 50 - 60 - 70 - 80 - 90 - 100 - 110 - 120 - 130
140 - 150 - 160 - 170 - 180 - 190 - 200 - 210 - 220 - 230 - 240 - 250 - 260
270 - 280 - 290 - 300 - 310 - 320 - 330 - 340 - 350 - 360 - 370 - 380 - 390
400 - 410 - 420 - 430 - 440 - 450 - 460 - 470 - 480 - 490 - 500 - 510 - 520
530 - 540 - 550 - 560 - 570 - 580 - 590 - 600 - 610 - 620 - 630 - 640 - 650
660 - 670 - 680 - 690 - 700 - 710 - 720 - 730 - 740 - 750 - 760 - 770 - 780
790 - 800 - 810 - 820 - 830 - 840 - 850 - 860 - 870 - 880 - 890 - 900 - 910
920 - 930 - 940 - 950 - 960 - 970 - 980 - 990 - 1000
Notes: ____

____

____

Reg Reps: 10 - 20 - 30 - 40 - 50 - 60 - 70 - 80 - 90 - 100 - 110 - 120 - 130
140 - 150 - 160 - 170 - 180 - 190 - 200 - 210 - 220 - 230 - 240 - 250 - 260
270 - 280 - 290 - 300 - 310 - 320 - 330 - 340 - 350 - 360 - 370 - 380 - 390
400 - 410 - 420 - 430 - 440 - 450 - 460 - 470 - 480 - 490 - 500 - 510 - 520
530 - 540 - 550 - 560 - 570 - 580 - 590 - 600 - 610 - 620 - 630 - 640 - 650
660 - 670 - 680 - 690 - 700 - 710 - 720 - 730 - 740 - 750 - 760 - 770 - 780
790 - 800 - 810 - 820 - 830 - 840 - 850 - 860 - 870 - 880 - 890 - 900 - 910
920 - 930 - 940 - 950 - 960 - 970 - 980 - 990 - 1000

Notes: ______________________________

Reg Reps: 10 - 20 - 30 - 40 - 50 - 60 - 70 - 80 - 90 - 100 - 110 - 120 - 130
140 - 150 - 160 - 170 - 180 - 190 - 200 - 210 - 220 - 230 - 240 - 250 - 260
270 - 280 - 290 - 300 - 310 - 320 - 330 - 340 - 350 - 360 - 370 - 380 - 390
400 - 410 - 420 - 430 - 440 - 450 - 460 - 470 - 480 - 490 - 500 - 510 - 520
530 - 540 - 550 - 560 - 570 - 580 - 590 - 600 - 610 - 620 - 630 - 640 - 650
660 - 670 - 680 - 690 - 700 - 710 - 720 - 730 - 740 - 750 - 760 - 770 - 780
790 - 800 - 810 - 820 - 830 - 840 - 850 - 860 - 870 - 880 - 890 - 900 - 910
920 - 930 - 940 - 950 - 960 - 970 - 980 - 990 - 1000

Notes: ______________________________

Reg Reps: 10 - 20 - 30 - 40 - 50 - 60 - 70 - 80 - 90 - 100 - 110 - 120 - 130
140 - 150 - 160 - 170 - 180 - 190 - 200 - 210 - 220 - 230 - 240 - 250 - 260
270 - 280 - 290 - 300 - 310 - 320 - 330 - 340 - 350 - 360 - 370 - 380 - 390
400 - 410 - 420 - 430 - 440 - 450 - 460 - 470 - 480 - 490 - 500 - 510 - 520
530 - 540 - 550 - 560 - 570 - 580 - 590 - 600 - 610 - 620 - 630 - 640 - 650
660 - 670 - 680 - 690 - 700 - 710 - 720 - 730 - 740 - 750 - 760 - 770 - 780
790 - 800 - 810 - 820 - 830 - 840 - 850 - 860 - 870 - 880 - 890 - 900 - 910
920 - 930 - 940 - 950 - 960 - 970 - 980 - 990 - 1000

Notes: ______________________________

End Date: __________

MAGAZINE CHANGE WHILE LYING ON YOUR BACK

Holster Position:______________ **Hand:** ______ **Start Date:** ________
Spare Magazine Position: ___________ **Firearm:** _________________
Cover Garment: _________________________________ **Caliber:** ______
Notes:____

____

____

Slow Reps: 10 - 20 - 30 - 40 - 50 - 60 - 70 - 80 - 90 - 100 - 110 - 120 - 130
140 - 150 - 160 - 170 - 180 - 190 - 200 - 210 - 220 - 230 - 240 - 250 - 260
270 - 280 - 290 - 300 - 310 - 320 - 330 - 340 - 350 - 360 - 370 - 380 - 390
400 - 410 - 420 - 430 - 440 - 450 - 460 - 470 - 480 - 490 - 500 - 510 - 520
530 - 540 - 550 - 560 - 570 - 580 - 590 - 600 - 610 - 620 - 630 - 640 - 650
660 - 670 - 680 - 690 - 700 - 710 - 720 - 730 - 740 - 750 - 760 - 770 - 780
790 - 800 - 810 - 820 - 830 - 840 - 850 - 860 - 870 - 880 - 890 - 900 - 910
920 - 930 - 940 - 950 - 960 - 970 - 980 - 990 - 1000

Reg Reps: 10 - 20 - 30 - 40 - 50 - 60 - 70 - 80 - 90 - 100 - 110 - 120 - 130
140 - 150 - 160 - 170 - 180 - 190 - 200 - 210 - 220 - 230 - 240 - 250 - 260
270 - 280 - 290 - 300 - 310 - 320 - 330 - 340 - 350 - 360 - 370 - 380 - 390
400 - 410 - 420 - 430 - 440 - 450 - 460 - 470 - 480 - 490 - 500 - 510 - 520
530 - 540 - 550 - 560 - 570 - 580 - 590 - 600 - 610 - 620 - 630 - 640 - 650
660 - 670 - 680 - 690 - 700 - 710 - 720 - 730 - 740 - 750 - 760 - 770 - 780
790 - 800 - 810 - 820 - 830 - 840 - 850 - 860 - 870 - 880 - 890 - 900 - 910
920 - 930 - 940 - 950 - 960 - 970 - 980 - 990 - 1000
Notes: ____

____

____

Reg Reps: 10 - 20 - 30 - 40 - 50 - 60 - 70 - 80 - 90 - 100 - 110 - 120 - 130
140 - 150 - 160 - 170 - 180 - 190 - 200 - 210 - 220 - 230 - 240 - 250 - 260
270 - 280 - 290 - 300 - 310 - 320 - 330 - 340 - 350 - 360 - 370 - 380 - 390
400 - 410 - 420 - 430 - 440 - 450 - 460 - 470 - 480 - 490 - 500 - 510 - 520
530 - 540 - 550 - 560 - 570 - 580 - 590 - 600 - 610 - 620 - 630 - 640 - 650
660 - 670 - 680 - 690 - 700 - 710 - 720 - 730 - 740 - 750 - 760 - 770 - 780
790 - 800 - 810 - 820 - 830 - 840 - 850 - 860 - 870 - 880 - 890 - 900 - 910
920 - 930 - 940 - 950 - 960 - 970 - 980 - 990 - 1000
Notes: ____

____

____

Reg Reps: 10 - 20 - 30 - 40 - 50 - 60 - 70 - 80 - 90 - 100 - 110 - 120 - 130 140 - 150 - 160 - 170 - 180 - 190 - 200 - 210 - 220 - 230 - 240 - 250 - 260 270 - 280 - 290 - 300 - 310 - 320 - 330 - 340 - 350 - 360 - 370 - 380 - 390 400 - 410 - 420 - 430 - 440 - 450 - 460 - 470 - 480 - 490 - 500 - 510 - 520 530 - 540 - 550 - 560 - 570 - 580 - 590 - 600 - 610 - 620 - 630 - 640 - 650 660 - 670 - 680 - 690 - 700 - 710 - 720 - 730 - 740 - 750 - 760 - 770 - 780 790 - 800 - 810 - 820 - 830 - 840 - 850 - 860 - 870 - 880 - 890 - 900 - 910 920 - 930 - 940 - 950 - 960 - 970 - 980 - 990 - 1000

Notes: ____

____

____

____

Reg Reps: 10 - 20 - 30 - 40 - 50 - 60 - 70 - 80 - 90 - 100 - 110 - 120 - 130 140 - 150 - 160 - 170 - 180 - 190 - 200 - 210 - 220 - 230 - 240 - 250 - 260 270 - 280 - 290 - 300 - 310 - 320 - 330 - 340 - 350 - 360 - 370 - 380 - 390 400 - 410 - 420 - 430 - 440 - 450 - 460 - 470 - 480 - 490 - 500 - 510 - 520 530 - 540 - 550 - 560 - 570 - 580 - 590 - 600 - 610 - 620 - 630 - 640 - 650 660 - 670 - 680 - 690 - 700 - 710 - 720 - 730 - 740 - 750 - 760 - 770 - 780 790 - 800 - 810 - 820 - 830 - 840 - 850 - 860 - 870 - 880 - 890 - 900 - 910 920 - 930 - 940 - 950 - 960 - 970 - 980 - 990 - 1000

Notes: ____

____

____

____

Reg Reps: 10 - 20 - 30 - 40 - 50 - 60 - 70 - 80 - 90 - 100 - 110 - 120 - 130 140 - 150 - 160 - 170 - 180 - 190 - 200 - 210 - 220 - 230 - 240 - 250 - 260 270 - 280 - 290 - 300 - 310 - 320 - 330 - 340 - 350 - 360 - 370 - 380 - 390 400 - 410 - 420 - 430 - 440 - 450 - 460 - 470 - 480 - 490 - 500 - 510 - 520 530 - 540 - 550 - 560 - 570 - 580 - 590 - 600 - 610 - 620 - 630 - 640 - 650 660 - 670 - 680 - 690 - 700 - 710 - 720 - 730 - 740 - 750 - 760 - 770 - 780 790 - 800 - 810 - 820 - 830 - 840 - 850 - 860 - 870 - 880 - 890 - 900 - 910 920 - 930 - 940 - 950 - 960 - 970 - 980 - 990 - 1000

Notes: ____

____

____

______________________________ **End Date:** ______________

TURN 180 DEGREES AND DRAW

Holster Position:_____________ **Hand:** ______ **Start Date:** _________
Spare Magazine Position: ___________ **Firearm:** ________________
Cover Garment: _______________________________ **Caliber:** ______
Notes:____
____
____

Slow Reps: 10 - 20 - 30 - 40 - 50 - 60 - 70 - 80 - 90 - 100 - 110 - 120 - 130
140 - 150 - 160 - 170 - 180 - 190 - 200 - 210 - 220 - 230 - 240 - 250 - 260
270 - 280 - 290 - 300 - 310 - 320 - 330 - 340 - 350 - 360 - 370 - 380 - 390
400 - 410 - 420 - 430 - 440 - 450 - 460 - 470 - 480 - 490 - 500 - 510 - 520
530 - 540 - 550 - 560 - 570 - 580 - 590 - 600 - 610 - 620 - 630 - 640 - 650
660 - 670 - 680 - 690 - 700 - 710 - 720 - 730 - 740 - 750 - 760 - 770 - 780
790 - 800 - 810 - 820 - 830 - 840 - 850 - 860 - 870 - 880 - 890 - 900 - 910
920 - 930 - 940 - 950 - 960 - 970 - 980 - 990 - 1000

Reg Reps: 10 - 20 - 30 - 40 - 50 - 60 - 70 - 80 - 90 - 100 - 110 - 120 - 130
140 - 150 - 160 - 170 - 180 - 190 - 200 - 210 - 220 - 230 - 240 - 250 - 260
270 - 280 - 290 - 300 - 310 - 320 - 330 - 340 - 350 - 360 - 370 - 380 - 390
400 - 410 - 420 - 430 - 440 - 450 - 460 - 470 - 480 - 490 - 500 - 510 - 520
530 - 540 - 550 - 560 - 570 - 580 - 590 - 600 - 610 - 620 - 630 - 640 - 650
660 - 670 - 680 - 690 - 700 - 710 - 720 - 730 - 740 - 750 - 760 - 770 - 780
790 - 800 - 810 - 820 - 830 - 840 - 850 - 860 - 870 - 880 - 890 - 900 - 910
920 - 930 - 940 - 950 - 960 - 970 - 980 - 990 - 1000
Notes: ____
____
____

Reg Reps: 10 - 20 - 30 - 40 - 50 - 60 - 70 - 80 - 90 - 100 - 110 - 120 - 130
140 - 150 - 160 - 170 - 180 - 190 - 200 - 210 - 220 - 230 - 240 - 250 - 260
270 - 280 - 290 - 300 - 310 - 320 - 330 - 340 - 350 - 360 - 370 - 380 - 390
400 - 410 - 420 - 430 - 440 - 450 - 460 - 470 - 480 - 490 - 500 - 510 - 520
530 - 540 - 550 - 560 - 570 - 580 - 590 - 600 - 610 - 620 - 630 - 640 - 650
660 - 670 - 680 - 690 - 700 - 710 - 720 - 730 - 740 - 750 - 760 - 770 - 780
790 - 800 - 810 - 820 - 830 - 840 - 850 - 860 - 870 - 880 - 890 - 900 - 910
920 - 930 - 940 - 950 - 960 - 970 - 980 - 990 - 1000
Notes: ____
____
____

Reg Reps: 10 - 20 - 30 - 40 - 50 - 60 - 70 - 80 - 90 - 100 - 110 - 120 - 130
140 - 150 - 160 - 170 - 180 - 190 - 200 - 210 - 220 - 230 - 240 - 250 - 260
270 - 280 - 290 - 300 - 310 - 320 - 330 - 340 - 350 - 360 - 370 - 380 - 390
400 - 410 - 420 - 430 - 440 - 450 - 460 - 470 - 480 - 490 - 500 - 510 - 520
530 - 540 - 550 - 560 - 570 - 580 - 590 - 600 - 610 - 620 - 630 - 640 - 650
660 - 670 - 680 - 690 - 700 - 710 - 720 - 730 - 740 - 750 - 760 - 770 - 780
790 - 800 - 810 - 820 - 830 - 840 - 850 - 860 - 870 - 880 - 890 - 900 - 910
920 - 930 - 940 - 950 - 960 - 970 - 980 - 990 - 1000

Notes: ____

____

____

____

Reg Reps: 10 - 20 - 30 - 40 - 50 - 60 - 70 - 80 - 90 - 100 - 110 - 120 - 130
140 - 150 - 160 - 170 - 180 - 190 - 200 - 210 - 220 - 230 - 240 - 250 - 260
270 - 280 - 290 - 300 - 310 - 320 - 330 - 340 - 350 - 360 - 370 - 380 - 390
400 - 410 - 420 - 430 - 440 - 450 - 460 - 470 - 480 - 490 - 500 - 510 - 520
530 - 540 - 550 - 560 - 570 - 580 - 590 - 600 - 610 - 620 - 630 - 640 - 650
660 - 670 - 680 - 690 - 700 - 710 - 720 - 730 - 740 - 750 - 760 - 770 - 780
790 - 800 - 810 - 820 - 830 - 840 - 850 - 860 - 870 - 880 - 890 - 900 - 910
920 - 930 - 940 - 950 - 960 - 970 - 980 - 990 - 1000

Notes: ____

____

____

____

Reg Reps: 10 - 20 - 30 - 40 - 50 - 60 - 70 - 80 - 90 - 100 - 110 - 120 - 130
140 - 150 - 160 - 170 - 180 - 190 - 200 - 210 - 220 - 230 - 240 - 250 - 260
270 - 280 - 290 - 300 - 310 - 320 - 330 - 340 - 350 - 360 - 370 - 380 - 390
400 - 410 - 420 - 430 - 440 - 450 - 460 - 470 - 480 - 490 - 500 - 510 - 520
530 - 540 - 550 - 560 - 570 - 580 - 590 - 600 - 610 - 620 - 630 - 640 - 650
660 - 670 - 680 - 690 - 700 - 710 - 720 - 730 - 740 - 750 - 760 - 770 - 780
790 - 800 - 810 - 820 - 830 - 840 - 850 - 860 - 870 - 880 - 890 - 900 - 910
920 - 930 - 940 - 950 - 960 - 970 - 980 - 990 - 1000

Notes: ____

____

____

____________________________ **End Date:** ____________

DRAW WHILE SEATED IN A VEHICLE: TO THE FRONT

Holster Position:______________ **Hand:** ______ **Start Date:** _________
Spare Magazine Position: ____________ **Firearm:** ________________
Cover Garment: ___________________________________ **Caliber:** ______
Notes:____

____

____

Slow Reps: 10 - 20 - 30 - 40 - 50 - 60 - 70 - 80 - 90 - 100 - 110 - 120 - 130
140 - 150 - 160 - 170 - 180 - 190 - 200 - 210 - 220 - 230 - 240 - 250 - 260
270 - 280 - 290 - 300 - 310 - 320 - 330 - 340 - 350 - 360 - 370 - 380 - 390
400 - 410 - 420 - 430 - 440 - 450 - 460 - 470 - 480 - 490 - 500 - 510 - 520
530 - 540 - 550 - 560 - 570 - 580 - 590 - 600 - 610 - 620 - 630 - 640 - 650
660 - 670 - 680 - 690 - 700 - 710 - 720 - 730 - 740 - 750 - 760 - 770 - 780
790 - 800 - 810 - 820 - 830 - 840 - 850 - 860 - 870 - 880 - 890 - 900 - 910
920 - 930 - 940 - 950 - 960 - 970 - 980 - 990 - 1000

Reg Reps: 10 - 20 - 30 - 40 - 50 - 60 - 70 - 80 - 90 - 100 - 110 - 120 - 130
140 - 150 - 160 - 170 - 180 - 190 - 200 - 210 - 220 - 230 - 240 - 250 - 260
270 - 280 - 290 - 300 - 310 - 320 - 330 - 340 - 350 - 360 - 370 - 380 - 390
400 - 410 - 420 - 430 - 440 - 450 - 460 - 470 - 480 - 490 - 500 - 510 - 520
530 - 540 - 550 - 560 - 570 - 580 - 590 - 600 - 610 - 620 - 630 - 640 - 650
660 - 670 - 680 - 690 - 700 - 710 - 720 - 730 - 740 - 750 - 760 - 770 - 780
790 - 800 - 810 - 820 - 830 - 840 - 850 - 860 - 870 - 880 - 890 - 900 - 910
920 - 930 - 940 - 950 - 960 - 970 - 980 - 990 - 1000
Notes: ____

____

____

Reg Reps: 10 - 20 - 30 - 40 - 50 - 60 - 70 - 80 - 90 - 100 - 110 - 120 - 130
140 - 150 - 160 - 170 - 180 - 190 - 200 - 210 - 220 - 230 - 240 - 250 - 260
270 - 280 - 290 - 300 - 310 - 320 - 330 - 340 - 350 - 360 - 370 - 380 - 390
400 - 410 - 420 - 430 - 440 - 450 - 460 - 470 - 480 - 490 - 500 - 510 - 520
530 - 540 - 550 - 560 - 570 - 580 - 590 - 600 - 610 - 620 - 630 - 640 - 650
660 - 670 - 680 - 690 - 700 - 710 - 720 - 730 - 740 - 750 - 760 - 770 - 780
790 - 800 - 810 - 820 - 830 - 840 - 850 - 860 - 870 - 880 - 890 - 900 - 910
920 - 930 - 940 - 950 - 960 - 970 - 980 - 990 - 1000
Notes: ____

____

____

Reg Reps: 10 - 20 - 30 - 40 - 50 - 60 - 70 - 80 - 90 - 100 - 110 - 120 - 130
140 - 150 - 160 - 170 - 180 - 190 - 200 - 210 - 220 - 230 - 240 - 250 - 260
270 - 280 - 290 - 300 - 310 - 320 - 330 - 340 - 350 - 360 - 370 - 380 - 390
400 - 410 - 420 - 430 - 440 - 450 - 460 - 470 - 480 - 490 - 500 - 510 - 520
530 - 540 - 550 - 560 - 570 - 580 - 590 - 600 - 610 - 620 - 630 - 640 - 650
660 - 670 - 680 - 690 - 700 - 710 - 720 - 730 - 740 - 750 - 760 - 770 - 780
790 - 800 - 810 - 820 - 830 - 840 - 850 - 860 - 870 - 880 - 890 - 900 - 910
920 - 930 - 940 - 950 - 960 - 970 - 980 - 990 - 1000

Notes: ______________________________

____

____

____

Reg Reps: 10 - 20 - 30 - 40 - 50 - 60 - 70 - 80 - 90 - 100 - 110 - 120 - 130
140 - 150 - 160 - 170 - 180 - 190 - 200 - 210 - 220 - 230 - 240 - 250 - 260
270 - 280 - 290 - 300 - 310 - 320 - 330 - 340 - 350 - 360 - 370 - 380 - 390
400 - 410 - 420 - 430 - 440 - 450 - 460 - 470 - 480 - 490 - 500 - 510 - 520
530 - 540 - 550 - 560 - 570 - 580 - 590 - 600 - 610 - 620 - 630 - 640 - 650
660 - 670 - 680 - 690 - 700 - 710 - 720 - 730 - 740 - 750 - 760 - 770 - 780
790 - 800 - 810 - 820 - 830 - 840 - 850 - 860 - 870 - 880 - 890 - 900 - 910
920 - 930 - 940 - 950 - 960 - 970 - 980 - 990 - 1000

Notes: ______________________________

____

____

____

Reg Reps: 10 - 20 - 30 - 40 - 50 - 60 - 70 - 80 - 90 - 100 - 110 - 120 - 130
140 - 150 - 160 - 170 - 180 - 190 - 200 - 210 - 220 - 230 - 240 - 250 - 260
270 - 280 - 290 - 300 - 310 - 320 - 330 - 340 - 350 - 360 - 370 - 380 - 390
400 - 410 - 420 - 430 - 440 - 450 - 460 - 470 - 480 - 490 - 500 - 510 - 520
530 - 540 - 550 - 560 - 570 - 580 - 590 - 600 - 610 - 620 - 630 - 640 - 650
660 - 670 - 680 - 690 - 700 - 710 - 720 - 730 - 740 - 750 - 760 - 770 - 780
790 - 800 - 810 - 820 - 830 - 840 - 850 - 860 - 870 - 880 - 890 - 900 - 910
920 - 930 - 940 - 950 - 960 - 970 - 980 - 990 - 1000

Notes: ______________________________

____

____

______________________________ **End Date:** ____________

DRAW WHILE SEATED IN A VEHICLE: TO THE LEFT

Holster Position:______________ **Hand:** ______ **Start Date:** _________
Spare Magazine Position: ____________ **Firearm:** ___________________
Cover Garment: ____________________________________ **Caliber:** ______
Notes:____

Slow Reps: 10 - 20 - 30 - 40 - 50 - 60 - 70 - 80 - 90 - 100 - 110 - 120 - 130
140 - 150 - 160 - 170 - 180 - 190 - 200 - 210 - 220 - 230 - 240 - 250 - 260
270 - 280 - 290 - 300 - 310 - 320 - 330 - 340 - 350 - 360 - 370 - 380 - 390
400 - 410 - 420 - 430 - 440 - 450 - 460 - 470 - 480 - 490 - 500 - 510 - 520
530 - 540 - 550 - 560 - 570 - 580 - 590 - 600 - 610 - 620 - 630 - 640 - 650
660 - 670 - 680 - 690 - 700 - 710 - 720 - 730 - 740 - 750 - 760 - 770 - 780
790 - 800 - 810 - 820 - 830 - 840 - 850 - 860 - 870 - 880 - 890 - 900 - 910
920 - 930 - 940 - 950 - 960 - 970 - 980 - 990 - 1000

Reg Reps: 10 - 20 - 30 - 40 - 50 - 60 - 70 - 80 - 90 - 100 - 110 - 120 - 130
140 - 150 - 160 - 170 - 180 - 190 - 200 - 210 - 220 - 230 - 240 - 250 - 260
270 - 280 - 290 - 300 - 310 - 320 - 330 - 340 - 350 - 360 - 370 - 380 - 390
400 - 410 - 420 - 430 - 440 - 450 - 460 - 470 - 480 - 490 - 500 - 510 - 520
530 - 540 - 550 - 560 - 570 - 580 - 590 - 600 - 610 - 620 - 630 - 640 - 650
660 - 670 - 680 - 690 - 700 - 710 - 720 - 730 - 740 - 750 - 760 - 770 - 780
790 - 800 - 810 - 820 - 830 - 840 - 850 - 860 - 870 - 880 - 890 - 900 - 910
920 - 930 - 940 - 950 - 960 - 970 - 980 - 990 - 1000
Notes: ____

Reg Reps: 10 - 20 - 30 - 40 - 50 - 60 - 70 - 80 - 90 - 100 - 110 - 120 - 130
140 - 150 - 160 - 170 - 180 - 190 - 200 - 210 - 220 - 230 - 240 - 250 - 260
270 - 280 - 290 - 300 - 310 - 320 - 330 - 340 - 350 - 360 - 370 - 380 - 390
400 - 410 - 420 - 430 - 440 - 450 - 460 - 470 - 480 - 490 - 500 - 510 - 520
530 - 540 - 550 - 560 - 570 - 580 - 590 - 600 - 610 - 620 - 630 - 640 - 650
660 - 670 - 680 - 690 - 700 - 710 - 720 - 730 - 740 - 750 - 760 - 770 - 780
790 - 800 - 810 - 820 - 830 - 840 - 850 - 860 - 870 - 880 - 890 - 900 - 910
920 - 930 - 940 - 950 - 960 - 970 - 980 - 990 - 1000
Notes: ____

Reg Reps: 10 - 20 - 30 - 40 - 50 - 60 - 70 - 80 - 90 - 100 - 110 - 120 - 130
140 - 150 - 160 - 170 - 180 - 190 - 200 - 210 - 220 - 230 - 240 - 250 - 260
270 - 280 - 290 - 300 - 310 - 320 - 330 - 340 - 350 - 360 - 370 - 380 - 390
400 - 410 - 420 - 430 - 440 - 450 - 460 - 470 - 480 - 490 - 500 - 510 - 520
530 - 540 - 550 - 560 - 570 - 580 - 590 - 600 - 610 - 620 - 630 - 640 - 650
660 - 670 - 680 - 690 - 700 - 710 - 720 - 730 - 740 - 750 - 760 - 770 - 780
790 - 800 - 810 - 820 - 830 - 840 - 850 - 860 - 870 - 880 - 890 - 900 - 910
920 - 930 - 940 - 950 - 960 - 970 - 980 - 990 - 1000

Notes: ____

____

____

____

Reg Reps: 10 - 20 - 30 - 40 - 50 - 60 - 70 - 80 - 90 - 100 - 110 - 120 - 130
140 - 150 - 160 - 170 - 180 - 190 - 200 - 210 - 220 - 230 - 240 - 250 - 260
270 - 280 - 290 - 300 - 310 - 320 - 330 - 340 - 350 - 360 - 370 - 380 - 390
400 - 410 - 420 - 430 - 440 - 450 - 460 - 470 - 480 - 490 - 500 - 510 - 520
530 - 540 - 550 - 560 - 570 - 580 - 590 - 600 - 610 - 620 - 630 - 640 - 650
660 - 670 - 680 - 690 - 700 - 710 - 720 - 730 - 740 - 750 - 760 - 770 - 780
790 - 800 - 810 - 820 - 830 - 840 - 850 - 860 - 870 - 880 - 890 - 900 - 910
920 - 930 - 940 - 950 - 960 - 970 - 980 - 990 - 1000

Notes: ____

____

____

____

Reg Reps: 10 - 20 - 30 - 40 - 50 - 60 - 70 - 80 - 90 - 100 - 110 - 120 - 130
140 - 150 - 160 - 170 - 180 - 190 - 200 - 210 - 220 - 230 - 240 - 250 - 260
270 - 280 - 290 - 300 - 310 - 320 - 330 - 340 - 350 - 360 - 370 - 380 - 390
400 - 410 - 420 - 430 - 440 - 450 - 460 - 470 - 480 - 490 - 500 - 510 - 520
530 - 540 - 550 - 560 - 570 - 580 - 590 - 600 - 610 - 620 - 630 - 640 - 650
660 - 670 - 680 - 690 - 700 - 710 - 720 - 730 - 740 - 750 - 760 - 770 - 780
790 - 800 - 810 - 820 - 830 - 840 - 850 - 860 - 870 - 880 - 890 - 900 - 910
920 - 930 - 940 - 950 - 960 - 970 - 980 - 990 - 1000

Notes: ____

____

____

______________________________________ **End Date:** __________

DRAW WHILE SEATED IN A VEHICLE: TO THE RIGHT

Holster Position:______________ **Hand:** ______ **Start Date:** _________
Spare Magazine Position: ___________ **Firearm:** __________________
Cover Garment: ________________________________ **Caliber:** ______
Notes:____

____

____

Slow Reps: 10 - 20 - 30 - 40 - 50 - 60 - 70 - 80 - 90 - 100 - 110 - 120 - 130
140 - 150 - 160 - 170 - 180 - 190 - 200 - 210 - 220 - 230 - 240 - 250 - 260
270 - 280 - 290 - 300 - 310 - 320 - 330 - 340 - 350 - 360 - 370 - 380 - 390
400 - 410 - 420 - 430 - 440 - 450 - 460 - 470 - 480 - 490 - 500 - 510 - 520
530 - 540 - 550 - 560 - 570 - 580 - 590 - 600 - 610 - 620 - 630 - 640 - 650
660 - 670 - 680 - 690 - 700 - 710 - 720 - 730 - 740 - 750 - 760 - 770 - 780
790 - 800 - 810 - 820 - 830 - 840 - 850 - 860 - 870 - 880 - 890 - 900 - 910
920 - 930 - 940 - 950 - 960 - 970 - 980 - 990 - 1000

Reg Reps: 10 - 20 - 30 - 40 - 50 - 60 - 70 - 80 - 90 - 100 - 110 - 120 - 130
140 - 150 - 160 - 170 - 180 - 190 - 200 - 210 - 220 - 230 - 240 - 250 - 260
270 - 280 - 290 - 300 - 310 - 320 - 330 - 340 - 350 - 360 - 370 - 380 - 390
400 - 410 - 420 - 430 - 440 - 450 - 460 - 470 - 480 - 490 - 500 - 510 - 520
530 - 540 - 550 - 560 - 570 - 580 - 590 - 600 - 610 - 620 - 630 - 640 - 650
660 - 670 - 680 - 690 - 700 - 710 - 720 - 730 - 740 - 750 - 760 - 770 - 780
790 - 800 - 810 - 820 - 830 - 840 - 850 - 860 - 870 - 880 - 890 - 900 - 910
920 - 930 - 940 - 950 - 960 - 970 - 980 - 990 - 1000
Notes: ____

____

____

Reg Reps: 10 - 20 - 30 - 40 - 50 - 60 - 70 - 80 - 90 - 100 - 110 - 120 - 130
140 - 150 - 160 - 170 - 180 - 190 - 200 - 210 - 220 - 230 - 240 - 250 - 260
270 - 280 - 290 - 300 - 310 - 320 - 330 - 340 - 350 - 360 - 370 - 380 - 390
400 - 410 - 420 - 430 - 440 - 450 - 460 - 470 - 480 - 490 - 500 - 510 - 520
530 - 540 - 550 - 560 - 570 - 580 - 590 - 600 - 610 - 620 - 630 - 640 - 650
660 - 670 - 680 - 690 - 700 - 710 - 720 - 730 - 740 - 750 - 760 - 770 - 780
790 - 800 - 810 - 820 - 830 - 840 - 850 - 860 - 870 - 880 - 890 - 900 - 910
920 - 930 - 940 - 950 - 960 - 970 - 980 - 990 - 1000
Notes: ____

____

____

Reg Reps: 10 - 20 - 30 - 40 - 50 - 60 - 70 - 80 - 90 - 100 - 110 - 120 - 130
140 - 150 - 160 - 170 - 180 - 190 - 200 - 210 - 220 - 230 - 240 - 250 - 260
270 - 280 - 290 - 300 - 310 - 320 - 330 - 340 - 350 - 360 - 370 - 380 - 390
400 - 410 - 420 - 430 - 440 - 450 - 460 - 470 - 480 - 490 - 500 - 510 - 520
530 - 540 - 550 - 560 - 570 - 580 - 590 - 600 - 610 - 620 - 630 - 640 - 650
660 - 670 - 680 - 690 - 700 - 710 - 720 - 730 - 740 - 750 - 760 - 770 - 780
790 - 800 - 810 - 820 - 830 - 840 - 850 - 860 - 870 - 880 - 890 - 900 - 910
920 - 930 - 940 - 950 - 960 - 970 - 980 - 990 - 1000

Notes: ______________________________

____

____

____

Reg Reps: 10 - 20 - 30 - 40 - 50 - 60 - 70 - 80 - 90 - 100 - 110 - 120 - 130
140 - 150 - 160 - 170 - 180 - 190 - 200 - 210 - 220 - 230 - 240 - 250 - 260
270 - 280 - 290 - 300 - 310 - 320 - 330 - 340 - 350 - 360 - 370 - 380 - 390
400 - 410 - 420 - 430 - 440 - 450 - 460 - 470 - 480 - 490 - 500 - 510 - 520
530 - 540 - 550 - 560 - 570 - 580 - 590 - 600 - 610 - 620 - 630 - 640 - 650
660 - 670 - 680 - 690 - 700 - 710 - 720 - 730 - 740 - 750 - 760 - 770 - 780
790 - 800 - 810 - 820 - 830 - 840 - 850 - 860 - 870 - 880 - 890 - 900 - 910
920 - 930 - 940 - 950 - 960 - 970 - 980 - 990 - 1000

Notes: ______________________________

____

____

____

Reg Reps: 10 - 20 - 30 - 40 - 50 - 60 - 70 - 80 - 90 - 100 - 110 - 120 - 130
140 - 150 - 160 - 170 - 180 - 190 - 200 - 210 - 220 - 230 - 240 - 250 - 260
270 - 280 - 290 - 300 - 310 - 320 - 330 - 340 - 350 - 360 - 370 - 380 - 390
400 - 410 - 420 - 430 - 440 - 450 - 460 - 470 - 480 - 490 - 500 - 510 - 520
530 - 540 - 550 - 560 - 570 - 580 - 590 - 600 - 610 - 620 - 630 - 640 - 650
660 - 670 - 680 - 690 - 700 - 710 - 720 - 730 - 740 - 750 - 760 - 770 - 780
790 - 800 - 810 - 820 - 830 - 840 - 850 - 860 - 870 - 880 - 890 - 900 - 910
920 - 930 - 940 - 950 - 960 - 970 - 980 - 990 - 1000

Notes: ______________________________

____

____

______________________________ **End Date:** __________

DRAW WHILE SEATED IN A VEHICLE: TO THE REAR

Holster Position:______________ **Hand:** ______ **Start Date:** _________
Spare Magazine Position: ___________ **Firearm:** __________________
Cover Garment: ________________________________ **Caliber:** ______
Notes:____

Slow Reps: 10 - 20 - 30 - 40 - 50 - 60 - 70 - 80 - 90 - 100 - 110 - 120 - 130
140 - 150 - 160 - 170 - 180 - 190 - 200 - 210 - 220 - 230 - 240 - 250 - 260
270 - 280 - 290 - 300 - 310 - 320 - 330 - 340 - 350 - 360 - 370 - 380 - 390
400 - 410 - 420 - 430 - 440 - 450 - 460 - 470 - 480 - 490 - 500 - 510 - 520
530 - 540 - 550 - 560 - 570 - 580 - 590 - 600 - 610 - 620 - 630 - 640 - 650
660 - 670 - 680 - 690 - 700 - 710 - 720 - 730 - 740 - 750 - 760 - 770 - 780
790 - 800 - 810 - 820 - 830 - 840 - 850 - 860 - 870 - 880 - 890 - 900 - 910
920 - 930 - 940 - 950 - 960 - 970 - 980 - 990 - 1000

Reg Reps: 10 - 20 - 30 - 40 - 50 - 60 - 70 - 80 - 90 - 100 - 110 - 120 - 130
140 - 150 - 160 - 170 - 180 - 190 - 200 - 210 - 220 - 230 - 240 - 250 - 260
270 - 280 - 290 - 300 - 310 - 320 - 330 - 340 - 350 - 360 - 370 - 380 - 390
400 - 410 - 420 - 430 - 440 - 450 - 460 - 470 - 480 - 490 - 500 - 510 - 520
530 - 540 - 550 - 560 - 570 - 580 - 590 - 600 - 610 - 620 - 630 - 640 - 650
660 - 670 - 680 - 690 - 700 - 710 - 720 - 730 - 740 - 750 - 760 - 770 - 780
790 - 800 - 810 - 820 - 830 - 840 - 850 - 860 - 870 - 880 - 890 - 900 - 910
920 - 930 - 940 - 950 - 960 - 970 - 980 - 990 - 1000
Notes: ____

Reg Reps: 10 - 20 - 30 - 40 - 50 - 60 - 70 - 80 - 90 - 100 - 110 - 120 - 130
140 - 150 - 160 - 170 - 180 - 190 - 200 - 210 - 220 - 230 - 240 - 250 - 260
270 - 280 - 290 - 300 - 310 - 320 - 330 - 340 - 350 - 360 - 370 - 380 - 390
400 - 410 - 420 - 430 - 440 - 450 - 460 - 470 - 480 - 490 - 500 - 510 - 520
530 - 540 - 550 - 560 - 570 - 580 - 590 - 600 - 610 - 620 - 630 - 640 - 650
660 - 670 - 680 - 690 - 700 - 710 - 720 - 730 - 740 - 750 - 760 - 770 - 780
790 - 800 - 810 - 820 - 830 - 840 - 850 - 860 - 870 - 880 - 890 - 900 - 910
920 - 930 - 940 - 950 - 960 - 970 - 980 - 990 - 1000
Notes: ____

Reg Reps: 10 - 20 - 30 - 40 - 50 - 60 - 70 - 80 - 90 - 100 - 110 - 120 - 130
140 - 150 - 160 - 170 - 180 - 190 - 200 - 210 - 220 - 230 - 240 - 250 - 260
270 - 280 - 290 - 300 - 310 - 320 - 330 - 340 - 350 - 360 - 370 - 380 - 390
400 - 410 - 420 - 430 - 440 - 450 - 460 - 470 - 480 - 490 - 500 - 510 - 520
530 - 540 - 550 - 560 - 570 - 580 - 590 - 600 - 610 - 620 - 630 - 640 - 650
660 - 670 - 680 - 690 - 700 - 710 - 720 - 730 - 740 - 750 - 760 - 770 - 780
790 - 800 - 810 - 820 - 830 - 840 - 850 - 860 - 870 - 880 - 890 - 900 - 910
920 - 930 - 940 - 950 - 960 - 970 - 980 - 990 - 1000

Notes: ______________________________

Reg Reps: 10 - 20 - 30 - 40 - 50 - 60 - 70 - 80 - 90 - 100 - 110 - 120 - 130
140 - 150 - 160 - 170 - 180 - 190 - 200 - 210 - 220 - 230 - 240 - 250 - 260
270 - 280 - 290 - 300 - 310 - 320 - 330 - 340 - 350 - 360 - 370 - 380 - 390
400 - 410 - 420 - 430 - 440 - 450 - 460 - 470 - 480 - 490 - 500 - 510 - 520
530 - 540 - 550 - 560 - 570 - 580 - 590 - 600 - 610 - 620 - 630 - 640 - 650
660 - 670 - 680 - 690 - 700 - 710 - 720 - 730 - 740 - 750 - 760 - 770 - 780
790 - 800 - 810 - 820 - 830 - 840 - 850 - 860 - 870 - 880 - 890 - 900 - 910
920 - 930 - 940 - 950 - 960 - 970 - 980 - 990 - 1000

Notes: ______________________________

Reg Reps: 10 - 20 - 30 - 40 - 50 - 60 - 70 - 80 - 90 - 100 - 110 - 120 - 130
140 - 150 - 160 - 170 - 180 - 190 - 200 - 210 - 220 - 230 - 240 - 250 - 260
270 - 280 - 290 - 300 - 310 - 320 - 330 - 340 - 350 - 360 - 370 - 380 - 390
400 - 410 - 420 - 430 - 440 - 450 - 460 - 470 - 480 - 490 - 500 - 510 - 520
530 - 540 - 550 - 560 - 570 - 580 - 590 - 600 - 610 - 620 - 630 - 640 - 650
660 - 670 - 680 - 690 - 700 - 710 - 720 - 730 - 740 - 750 - 760 - 770 - 780
790 - 800 - 810 - 820 - 830 - 840 - 850 - 860 - 870 - 880 - 890 - 900 - 910
920 - 930 - 940 - 950 - 960 - 970 - 980 - 990 - 1000

Notes: ______________________________

End Date: ____________

DRAW WHILE CARRYING SOMETHING IN YOUR HANDS

Holster Position:______________ **Hand:** ______ **Start Date:** _________
Spare Magazine Position: ____________ **Firearm:** __________________
Cover Garment: __________________________________ **Caliber:** ______
Notes:____

____

____

Slow Reps: 10 - 20 - 30 - 40 - 50 - 60 - 70 - 80 - 90 - 100 - 110 - 120 - 130
140 - 150 - 160 - 170 - 180 - 190 - 200 - 210 - 220 - 230 - 240 - 250 - 260
270 - 280 - 290 - 300 - 310 - 320 - 330 - 340 - 350 - 360 - 370 - 380 - 390
400 - 410 - 420 - 430 - 440 - 450 - 460 - 470 - 480 - 490 - 500 - 510 - 520
530 - 540 - 550 - 560 - 570 - 580 - 590 - 600 - 610 - 620 - 630 - 640 - 650
660 - 670 - 680 - 690 - 700 - 710 - 720 - 730 - 740 - 750 - 760 - 770 - 780
790 - 800 - 810 - 820 - 830 - 840 - 850 - 860 - 870 - 880 - 890 - 900 - 910
920 - 930 - 940 - 950 - 960 - 970 - 980 - 990 - 1000

Reg Reps: 10 - 20 - 30 - 40 - 50 - 60 - 70 - 80 - 90 - 100 - 110 - 120 - 130
140 - 150 - 160 - 170 - 180 - 190 - 200 - 210 - 220 - 230 - 240 - 250 - 260
270 - 280 - 290 - 300 - 310 - 320 - 330 - 340 - 350 - 360 - 370 - 380 - 390
400 - 410 - 420 - 430 - 440 - 450 - 460 - 470 - 480 - 490 - 500 - 510 - 520
530 - 540 - 550 - 560 - 570 - 580 - 590 - 600 - 610 - 620 - 630 - 640 - 650
660 - 670 - 680 - 690 - 700 - 710 - 720 - 730 - 740 - 750 - 760 - 770 - 780
790 - 800 - 810 - 820 - 830 - 840 - 850 - 860 - 870 - 880 - 890 - 900 - 910
920 - 930 - 940 - 950 - 960 - 970 - 980 - 990 - 1000
Notes: ____

____

____

Reg Reps: 10 - 20 - 30 - 40 - 50 - 60 - 70 - 80 - 90 - 100 - 110 - 120 - 130
140 - 150 - 160 - 170 - 180 - 190 - 200 - 210 - 220 - 230 - 240 - 250 - 260
270 - 280 - 290 - 300 - 310 - 320 - 330 - 340 - 350 - 360 - 370 - 380 - 390
400 - 410 - 420 - 430 - 440 - 450 - 460 - 470 - 480 - 490 - 500 - 510 - 520
530 - 540 - 550 - 560 - 570 - 580 - 590 - 600 - 610 - 620 - 630 - 640 - 650
660 - 670 - 680 - 690 - 700 - 710 - 720 - 730 - 740 - 750 - 760 - 770 - 780
790 - 800 - 810 - 820 - 830 - 840 - 850 - 860 - 870 - 880 - 890 - 900 - 910
920 - 930 - 940 - 950 - 960 - 970 - 980 - 990 - 1000
Notes: ____

____

____

Reg Reps: 10 - 20 - 30 - 40 - 50 - 60 - 70 - 80 - 90 - 100 - 110 - 120 - 130 - 140 - 150 - 160 - 170 - 180 - 190 - 200 - 210 - 220 - 230 - 240 - 250 - 260 - 270 - 280 - 290 - 300 - 310 - 320 - 330 - 340 - 350 - 360 - 370 - 380 - 390 - 400 - 410 - 420 - 430 - 440 - 450 - 460 - 470 - 480 - 490 - 500 - 510 - 520 - 530 - 540 - 550 - 560 - 570 - 580 - 590 - 600 - 610 - 620 - 630 - 640 - 650 - 660 - 670 - 680 - 690 - 700 - 710 - 720 - 730 - 740 - 750 - 760 - 770 - 780 - 790 - 800 - 810 - 820 - 830 - 840 - 850 - 860 - 870 - 880 - 890 - 900 - 910 - 920 - 930 - 940 - 950 - 960 - 970 - 980 - 990 - 1000

Notes: ____

____

____

____

Reg Reps: 10 - 20 - 30 - 40 - 50 - 60 - 70 - 80 - 90 - 100 - 110 - 120 - 130 - 140 - 150 - 160 - 170 - 180 - 190 - 200 - 210 - 220 - 230 - 240 - 250 - 260 - 270 - 280 - 290 - 300 - 310 - 320 - 330 - 340 - 350 - 360 - 370 - 380 - 390 - 400 - 410 - 420 - 430 - 440 - 450 - 460 - 470 - 480 - 490 - 500 - 510 - 520 - 530 - 540 - 550 - 560 - 570 - 580 - 590 - 600 - 610 - 620 - 630 - 640 - 650 - 660 - 670 - 680 - 690 - 700 - 710 - 720 - 730 - 740 - 750 - 760 - 770 - 780 - 790 - 800 - 810 - 820 - 830 - 840 - 850 - 860 - 870 - 880 - 890 - 900 - 910 - 920 - 930 - 940 - 950 - 960 - 970 - 980 - 990 - 1000

Notes: ____

____

____

____

Reg Reps: 10 - 20 - 30 - 40 - 50 - 60 - 70 - 80 - 90 - 100 - 110 - 120 - 130 - 140 - 150 - 160 - 170 - 180 - 190 - 200 - 210 - 220 - 230 - 240 - 250 - 260 - 270 - 280 - 290 - 300 - 310 - 320 - 330 - 340 - 350 - 360 - 370 - 380 - 390 - 400 - 410 - 420 - 430 - 440 - 450 - 460 - 470 - 480 - 490 - 500 - 510 - 520 - 530 - 540 - 550 - 560 - 570 - 580 - 590 - 600 - 610 - 620 - 630 - 640 - 650 - 660 - 670 - 680 - 690 - 700 - 710 - 720 - 730 - 740 - 750 - 760 - 770 - 780 - 790 - 800 - 810 - 820 - 830 - 840 - 850 - 860 - 870 - 880 - 890 - 900 - 910 - 920 - 930 - 940 - 950 - 960 - 970 - 980 - 990 - 1000

Notes: ____

____

____

____________________________ **End Date:** ______________

DRAW WITH YOUR NON-DOMINANT HAND

Holster Position:_____________ **Hand:** ______ **Start Date:** _________

Spare Magazine Position: ___________ **Firearm:** __________________

Cover Garment: _________________________________ **Caliber:** ______

Notes:____

____

____

Slow Reps: 10 - 20 - 30 - 40 - 50 - 60 - 70 - 80 - 90 - 100 - 110 - 120 - 130
140 - 150 - 160 - 170 - 180 - 190 - 200 - 210 - 220 - 230 - 240 - 250 - 260
270 - 280 - 290 - 300 - 310 - 320 - 330 - 340 - 350 - 360 - 370 - 380 - 390
400 - 410 - 420 - 430 - 440 - 450 - 460 - 470 - 480 - 490 - 500 - 510 - 520
530 - 540 - 550 - 560 - 570 - 580 - 590 - 600 - 610 - 620 - 630 - 640 - 650
660 - 670 - 680 - 690 - 700 - 710 - 720 - 730 - 740 - 750 - 760 - 770 - 780
790 - 800 - 810 - 820 - 830 - 840 - 850 - 860 - 870 - 880 - 890 - 900 - 910
920 - 930 - 940 - 950 - 960 - 970 - 980 - 990 - 1000

Reg Reps: 10 - 20 - 30 - 40 - 50 - 60 - 70 - 80 - 90 - 100 - 110 - 120 - 130
140 - 150 - 160 - 170 - 180 - 190 - 200 - 210 - 220 - 230 - 240 - 250 - 260
270 - 280 - 290 - 300 - 310 - 320 - 330 - 340 - 350 - 360 - 370 - 380 - 390
400 - 410 - 420 - 430 - 440 - 450 - 460 - 470 - 480 - 490 - 500 - 510 - 520
530 - 540 - 550 - 560 - 570 - 580 - 590 - 600 - 610 - 620 - 630 - 640 - 650
660 - 670 - 680 - 690 - 700 - 710 - 720 - 730 - 740 - 750 - 760 - 770 - 780
790 - 800 - 810 - 820 - 830 - 840 - 850 - 860 - 870 - 880 - 890 - 900 - 910
920 - 930 - 940 - 950 - 960 - 970 - 980 - 990 - 1000

Notes: ____

____

____

Reg Reps: 10 - 20 - 30 - 40 - 50 - 60 - 70 - 80 - 90 - 100 - 110 - 120 - 130
140 - 150 - 160 - 170 - 180 - 190 - 200 - 210 - 220 - 230 - 240 - 250 - 260
270 - 280 - 290 - 300 - 310 - 320 - 330 - 340 - 350 - 360 - 370 - 380 - 390
400 - 410 - 420 - 430 - 440 - 450 - 460 - 470 - 480 - 490 - 500 - 510 - 520
530 - 540 - 550 - 560 - 570 - 580 - 590 - 600 - 610 - 620 - 630 - 640 - 650
660 - 670 - 680 - 690 - 700 - 710 - 720 - 730 - 740 - 750 - 760 - 770 - 780
790 - 800 - 810 - 820 - 830 - 840 - 850 - 860 - 870 - 880 - 890 - 900 - 910
920 - 930 - 940 - 950 - 960 - 970 - 980 - 990 - 1000

Notes: ____

____

____

Reg Reps: 10 - 20 - 30 - 40 - 50 - 60 - 70 - 80 - 90 - 100 - 110 - 120 - 130
140 - 150 - 160 - 170 - 180 - 190 - 200 - 210 - 220 - 230 - 240 - 250 - 260
270 - 280 - 290 - 300 - 310 - 320 - 330 - 340 - 350 - 360 - 370 - 380 - 390
400 - 410 - 420 - 430 - 440 - 450 - 460 - 470 - 480 - 490 - 500 - 510 - 520
530 - 540 - 550 - 560 - 570 - 580 - 590 - 600 - 610 - 620 - 630 - 640 - 650
660 - 670 - 680 - 690 - 700 - 710 - 720 - 730 - 740 - 750 - 760 - 770 - 780
790 - 800 - 810 - 820 - 830 - 840 - 850 - 860 - 870 - 880 - 890 - 900 - 910
920 - 930 - 940 - 950 - 960 - 970 - 980 - 990 - 1000

Notes: ______________________________

____

____

____

Reg Reps: 10 - 20 - 30 - 40 - 50 - 60 - 70 - 80 - 90 - 100 - 110 - 120 - 130
140 - 150 - 160 - 170 - 180 - 190 - 200 - 210 - 220 - 230 - 240 - 250 - 260
270 - 280 - 290 - 300 - 310 - 320 - 330 - 340 - 350 - 360 - 370 - 380 - 390
400 - 410 - 420 - 430 - 440 - 450 - 460 - 470 - 480 - 490 - 500 - 510 - 520
530 - 540 - 550 - 560 - 570 - 580 - 590 - 600 - 610 - 620 - 630 - 640 - 650
660 - 670 - 680 - 690 - 700 - 710 - 720 - 730 - 740 - 750 - 760 - 770 - 780
790 - 800 - 810 - 820 - 830 - 840 - 850 - 860 - 870 - 880 - 890 - 900 - 910
920 - 930 - 940 - 950 - 960 - 970 - 980 - 990 - 1000

Notes: ______________________________

____

____

____

Reg Reps: 10 - 20 - 30 - 40 - 50 - 60 - 70 - 80 - 90 - 100 - 110 - 120 - 130
140 - 150 - 160 - 170 - 180 - 190 - 200 - 210 - 220 - 230 - 240 - 250 - 260
270 - 280 - 290 - 300 - 310 - 320 - 330 - 340 - 350 - 360 - 370 - 380 - 390
400 - 410 - 420 - 430 - 440 - 450 - 460 - 470 - 480 - 490 - 500 - 510 - 520
530 - 540 - 550 - 560 - 570 - 580 - 590 - 600 - 610 - 620 - 630 - 640 - 650
660 - 670 - 680 - 690 - 700 - 710 - 720 - 730 - 740 - 750 - 760 - 770 - 780
790 - 800 - 810 - 820 - 830 - 840 - 850 - 860 - 870 - 880 - 890 - 900 - 910
920 - 930 - 940 - 950 - 960 - 970 - 980 - 990 - 1000

Notes: ______________________________

____

____

______________________________ **End Date:** __________

MAGAZINE CHANGE: NON-DOMINANT HAND ONLY

Holster Position:____________ **Hand:** ______ **Start Date:** _________
Spare Magazine Position: ___________ **Firearm:** _________________
Cover Garment: _______________________________ **Caliber:** ______
Notes:____

____

Slow Reps: 10 - 20 - 30 - 40 - 50 - 60 - 70 - 80 - 90 - 100 - 110 - 120 - 130
140 - 150 - 160 - 170 - 180 - 190 - 200 - 210 - 220 - 230 - 240 - 250 - 260
270 - 280 - 290 - 300 - 310 - 320 - 330 - 340 - 350 - 360 - 370 - 380 - 390
400 - 410 - 420 - 430 - 440 - 450 - 460 - 470 - 480 - 490 - 500 - 510 - 520
530 - 540 - 550 - 560 - 570 - 580 - 590 - 600 - 610 - 620 - 630 - 640 - 650
660 - 670 - 680 - 690 - 700 - 710 - 720 - 730 - 740 - 750 - 760 - 770 - 780
790 - 800 - 810 - 820 - 830 - 840 - 850 - 860 - 870 - 880 - 890 - 900 - 910
920 - 930 - 940 - 950 - 960 - 970 - 980 - 990 - 1000

Reg Reps: 10 - 20 - 30 - 40 - 50 - 60 - 70 - 80 - 90 - 100 - 110 - 120 - 130
140 - 150 - 160 - 170 - 180 - 190 - 200 - 210 - 220 - 230 - 240 - 250 - 260
270 - 280 - 290 - 300 - 310 - 320 - 330 - 340 - 350 - 360 - 370 - 380 - 390
400 - 410 - 420 - 430 - 440 - 450 - 460 - 470 - 480 - 490 - 500 - 510 - 520
530 - 540 - 550 - 560 - 570 - 580 - 590 - 600 - 610 - 620 - 630 - 640 - 650
660 - 670 - 680 - 690 - 700 - 710 - 720 - 730 - 740 - 750 - 760 - 770 - 780
790 - 800 - 810 - 820 - 830 - 840 - 850 - 860 - 870 - 880 - 890 - 900 - 910
920 - 930 - 940 - 950 - 960 - 970 - 980 - 990 - 1000
Notes: ____

____

____

Reg Reps: 10 - 20 - 30 - 40 - 50 - 60 - 70 - 80 - 90 - 100 - 110 - 120 - 130
140 - 150 - 160 - 170 - 180 - 190 - 200 - 210 - 220 - 230 - 240 - 250 - 260
270 - 280 - 290 - 300 - 310 - 320 - 330 - 340 - 350 - 360 - 370 - 380 - 390
400 - 410 - 420 - 430 - 440 - 450 - 460 - 470 - 480 - 490 - 500 - 510 - 520
530 - 540 - 550 - 560 - 570 - 580 - 590 - 600 - 610 - 620 - 630 - 640 - 650
660 - 670 - 680 - 690 - 700 - 710 - 720 - 730 - 740 - 750 - 760 - 770 - 780
790 - 800 - 810 - 820 - 830 - 840 - 850 - 860 - 870 - 880 - 890 - 900 - 910
920 - 930 - 940 - 950 - 960 - 970 - 980 - 990 - 1000
Notes: ____

____

____

Reg Reps: 10 - 20 - 30 - 40 - 50 - 60 - 70 - 80 - 90 - 100 - 110 - 120 - 130 140 - 150 - 160 - 170 - 180 - 190 - 200 - 210 - 220 - 230 - 240 - 250 - 260 270 - 280 - 290 - 300 - 310 - 320 - 330 - 340 - 350 - 360 - 370 - 380 - 390 400 - 410 - 420 - 430 - 440 - 450 - 460 - 470 - 480 - 490 - 500 - 510 - 520 530 - 540 - 550 - 560 - 570 - 580 - 590 - 600 - 610 - 620 - 630 - 640 - 650 660 - 670 - 680 - 690 - 700 - 710 - 720 - 730 - 740 - 750 - 760 - 770 - 780 790 - 800 - 810 - 820 - 830 - 840 - 850 - 860 - 870 - 880 - 890 - 900 - 910 920 - 930 - 940 - 950 - 960 - 970 - 980 - 990 - 1000

Notes: ______________________________

____

____

____

Reg Reps: 10 - 20 - 30 - 40 - 50 - 60 - 70 - 80 - 90 - 100 - 110 - 120 - 130 140 - 150 - 160 - 170 - 180 - 190 - 200 - 210 - 220 - 230 - 240 - 250 - 260 270 - 280 - 290 - 300 - 310 - 320 - 330 - 340 - 350 - 360 - 370 - 380 - 390 400 - 410 - 420 - 430 - 440 - 450 - 460 - 470 - 480 - 490 - 500 - 510 - 520 530 - 540 - 550 - 560 - 570 - 580 - 590 - 600 - 610 - 620 - 630 - 640 - 650 660 - 670 - 680 - 690 - 700 - 710 - 720 - 730 - 740 - 750 - 760 - 770 - 780 790 - 800 - 810 - 820 - 830 - 840 - 850 - 860 - 870 - 880 - 890 - 900 - 910 920 - 930 - 940 - 950 - 960 - 970 - 980 - 990 - 1000

Notes: ______________________________

____

____

____

Reg Reps: 10 - 20 - 30 - 40 - 50 - 60 - 70 - 80 - 90 - 100 - 110 - 120 - 130 140 - 150 - 160 - 170 - 180 - 190 - 200 - 210 - 220 - 230 - 240 - 250 - 260 270 - 280 - 290 - 300 - 310 - 320 - 330 - 340 - 350 - 360 - 370 - 380 - 390 400 - 410 - 420 - 430 - 440 - 450 - 460 - 470 - 480 - 490 - 500 - 510 - 520 530 - 540 - 550 - 560 - 570 - 580 - 590 - 600 - 610 - 620 - 630 - 640 - 650 660 - 670 - 680 - 690 - 700 - 710 - 720 - 730 - 740 - 750 - 760 - 770 - 780 790 - 800 - 810 - 820 - 830 - 840 - 850 - 860 - 870 - 880 - 890 - 900 - 910 920 - 930 - 940 - 950 - 960 - 970 - 980 - 990 - 1000

Notes: ______________________________

____

____

______________________________ **End Date:** __________

TRANSITIONING FROM DOMINANT TO NON-DOMINANT HAND

Holster Position:_____________ **Hand:** ______ **Start Date:** _________
Spare Magazine Position: ___________ **Firearm:** __________________
Cover Garment: ________________________________ **Caliber:** ______
Notes:____

____

____

Slow Reps: 10 - 20 - 30 - 40 - 50 - 60 - 70 - 80 - 90 - 100 - 110 - 120 - 130
140 - 150 - 160 - 170 - 180 - 190 - 200 - 210 - 220 - 230 - 240 - 250 - 260
270 - 280 - 290 - 300 - 310 - 320 - 330 - 340 - 350 - 360 - 370 - 380 - 390
400 - 410 - 420 - 430 - 440 - 450 - 460 - 470 - 480 - 490 - 500 - 510 - 520
530 - 540 - 550 - 560 - 570 - 580 - 590 - 600 - 610 - 620 - 630 - 640 - 650
660 - 670 - 680 - 690 - 700 - 710 - 720 - 730 - 740 - 750 - 760 - 770 - 780
790 - 800 - 810 - 820 - 830 - 840 - 850 - 860 - 870 - 880 - 890 - 900 - 910
920 - 930 - 940 - 950 - 960 - 970 - 980 - 990 - 1000

Reg Reps: 10 - 20 - 30 - 40 - 50 - 60 - 70 - 80 - 90 - 100 - 110 - 120 - 130
140 - 150 - 160 - 170 - 180 - 190 - 200 - 210 - 220 - 230 - 240 - 250 - 260
270 - 280 - 290 - 300 - 310 - 320 - 330 - 340 - 350 - 360 - 370 - 380 - 390
400 - 410 - 420 - 430 - 440 - 450 - 460 - 470 - 480 - 490 - 500 - 510 - 520
530 - 540 - 550 - 560 - 570 - 580 - 590 - 600 - 610 - 620 - 630 - 640 - 650
660 - 670 - 680 - 690 - 700 - 710 - 720 - 730 - 740 - 750 - 760 - 770 - 780
790 - 800 - 810 - 820 - 830 - 840 - 850 - 860 - 870 - 880 - 890 - 900 - 910
920 - 930 - 940 - 950 - 960 - 970 - 980 - 990 - 1000
Notes: ____

____

____

Reg Reps: 10 - 20 - 30 - 40 - 50 - 60 - 70 - 80 - 90 - 100 - 110 - 120 - 130
140 - 150 - 160 - 170 - 180 - 190 - 200 - 210 - 220 - 230 - 240 - 250 - 260
270 - 280 - 290 - 300 - 310 - 320 - 330 - 340 - 350 - 360 - 370 - 380 - 390
400 - 410 - 420 - 430 - 440 - 450 - 460 - 470 - 480 - 490 - 500 - 510 - 520
530 - 540 - 550 - 560 - 570 - 580 - 590 - 600 - 610 - 620 - 630 - 640 - 650
660 - 670 - 680 - 690 - 700 - 710 - 720 - 730 - 740 - 750 - 760 - 770 - 780
790 - 800 - 810 - 820 - 830 - 840 - 850 - 860 - 870 - 880 - 890 - 900 - 910
920 - 930 - 940 - 950 - 960 - 970 - 980 - 990 - 1000
Notes: ____

____

____

Reg Reps: 10 - 20 - 30 - 40 - 50 - 60 - 70 - 80 - 90 - 100 - 110 - 120 - 130 - 140 - 150 - 160 - 170 - 180 - 190 - 200 - 210 - 220 - 230 - 240 - 250 - 260 - 270 - 280 - 290 - 300 - 310 - 320 - 330 - 340 - 350 - 360 - 370 - 380 - 390 - 400 - 410 - 420 - 430 - 440 - 450 - 460 - 470 - 480 - 490 - 500 - 510 - 520 - 530 - 540 - 550 - 560 - 570 - 580 - 590 - 600 - 610 - 620 - 630 - 640 - 650 - 660 - 670 - 680 - 690 - 700 - 710 - 720 - 730 - 740 - 750 - 760 - 770 - 780 - 790 - 800 - 810 - 820 - 830 - 840 - 850 - 860 - 870 - 880 - 890 - 900 - 910 - 920 - 930 - 940 - 950 - 960 - 970 - 980 - 990 - 1000

Notes: ______________________________

____

____

____

Reg Reps: 10 - 20 - 30 - 40 - 50 - 60 - 70 - 80 - 90 - 100 - 110 - 120 - 130 - 140 - 150 - 160 - 170 - 180 - 190 - 200 - 210 - 220 - 230 - 240 - 250 - 260 - 270 - 280 - 290 - 300 - 310 - 320 - 330 - 340 - 350 - 360 - 370 - 380 - 390 - 400 - 410 - 420 - 430 - 440 - 450 - 460 - 470 - 480 - 490 - 500 - 510 - 520 - 530 - 540 - 550 - 560 - 570 - 580 - 590 - 600 - 610 - 620 - 630 - 640 - 650 - 660 - 670 - 680 - 690 - 700 - 710 - 720 - 730 - 740 - 750 - 760 - 770 - 780 - 790 - 800 - 810 - 820 - 830 - 840 - 850 - 860 - 870 - 880 - 890 - 900 - 910 - 920 - 930 - 940 - 950 - 960 - 970 - 980 - 990 - 1000

Notes: ______________________________

____

____

____

Reg Reps: 10 - 20 - 30 - 40 - 50 - 60 - 70 - 80 - 90 - 100 - 110 - 120 - 130 - 140 - 150 - 160 - 170 - 180 - 190 - 200 - 210 - 220 - 230 - 240 - 250 - 260 - 270 - 280 - 290 - 300 - 310 - 320 - 330 - 340 - 350 - 360 - 370 - 380 - 390 - 400 - 410 - 420 - 430 - 440 - 450 - 460 - 470 - 480 - 490 - 500 - 510 - 520 - 530 - 540 - 550 - 560 - 570 - 580 - 590 - 600 - 610 - 620 - 630 - 640 - 650 - 660 - 670 - 680 - 690 - 700 - 710 - 720 - 730 - 740 - 750 - 760 - 770 - 780 - 790 - 800 - 810 - 820 - 830 - 840 - 850 - 860 - 870 - 880 - 890 - 900 - 910 - 920 - 930 - 940 - 950 - 960 - 970 - 980 - 990 - 1000

Notes: ______________________________

____

____

______________________________ **End Date:** __________

PICK HANDGUN UP FROM THE GROUND: DOMINANT HAND ONLY

Holster Position:______________ **Hand:** ______ **Start Date:** ________
Spare Magazine Position: ___________ **Firearm:** __________________
Cover Garment: ________________________________ **Caliber:** ______
Notes:____

____

____

Slow Reps: 10 - 20 - 30 - 40 - 50 - 60 - 70 - 80 - 90 - 100 - 110 - 120 - 130
140 - 150 - 160 - 170 - 180 - 190 - 200 - 210 - 220 - 230 - 240 - 250 - 260
270 - 280 - 290 - 300 - 310 - 320 - 330 - 340 - 350 - 360 - 370 - 380 - 390
400 - 410 - 420 - 430 - 440 - 450 - 460 - 470 - 480 - 490 - 500 - 510 - 520
530 - 540 - 550 - 560 - 570 - 580 - 590 - 600 - 610 - 620 - 630 - 640 - 650
660 - 670 - 680 - 690 - 700 - 710 - 720 - 730 - 740 - 750 - 760 - 770 - 780
790 - 800 - 810 - 820 - 830 - 840 - 850 - 860 - 870 - 880 - 890 - 900 - 910
920 - 930 - 940 - 950 - 960 - 970 - 980 - 990 - 1000

Reg Reps: 10 - 20 - 30 - 40 - 50 - 60 - 70 - 80 - 90 - 100 - 110 - 120 - 130
140 - 150 - 160 - 170 - 180 - 190 - 200 - 210 - 220 - 230 - 240 - 250 - 260
270 - 280 - 290 - 300 - 310 - 320 - 330 - 340 - 350 - 360 - 370 - 380 - 390
400 - 410 - 420 - 430 - 440 - 450 - 460 - 470 - 480 - 490 - 500 - 510 - 520
530 - 540 - 550 - 560 - 570 - 580 - 590 - 600 - 610 - 620 - 630 - 640 - 650
660 - 670 - 680 - 690 - 700 - 710 - 720 - 730 - 740 - 750 - 760 - 770 - 780
790 - 800 - 810 - 820 - 830 - 840 - 850 - 860 - 870 - 880 - 890 - 900 - 910
920 - 930 - 940 - 950 - 960 - 970 - 980 - 990 - 1000
Notes: ____

____

____

Reg Reps: 10 - 20 - 30 - 40 - 50 - 60 - 70 - 80 - 90 - 100 - 110 - 120 - 130
140 - 150 - 160 - 170 - 180 - 190 - 200 - 210 - 220 - 230 - 240 - 250 - 260
270 - 280 - 290 - 300 - 310 - 320 - 330 - 340 - 350 - 360 - 370 - 380 - 390
400 - 410 - 420 - 430 - 440 - 450 - 460 - 470 - 480 - 490 - 500 - 510 - 520
530 - 540 - 550 - 560 - 570 - 580 - 590 - 600 - 610 - 620 - 630 - 640 - 650
660 - 670 - 680 - 690 - 700 - 710 - 720 - 730 - 740 - 750 - 760 - 770 - 780
790 - 800 - 810 - 820 - 830 - 840 - 850 - 860 - 870 - 880 - 890 - 900 - 910
920 - 930 - 940 - 950 - 960 - 970 - 980 - 990 - 1000
Notes: ____

____

____

Reg Reps: 10 - 20 - 30 - 40 - 50 - 60 - 70 - 80 - 90 - 100 - 110 - 120 - 130
140 - 150 - 160 - 170 - 180 - 190 - 200 - 210 - 220 - 230 - 240 - 250 - 260
270 - 280 - 290 - 300 - 310 - 320 - 330 - 340 - 350 - 360 - 370 - 380 - 390
400 - 410 - 420 - 430 - 440 - 450 - 460 - 470 - 480 - 490 - 500 - 510 - 520
530 - 540 - 550 - 560 - 570 - 580 - 590 - 600 - 610 - 620 - 630 - 640 - 650
660 - 670 - 680 - 690 - 700 - 710 - 720 - 730 - 740 - 750 - 760 - 770 - 780
790 - 800 - 810 - 820 - 830 - 840 - 850 - 860 - 870 - 880 - 890 - 900 - 910
920 - 930 - 940 - 950 - 960 - 970 - 980 - 990 - 1000

Notes: ______________________________

____

____

____

Reg Reps: 10 - 20 - 30 - 40 - 50 - 60 - 70 - 80 - 90 - 100 - 110 - 120 - 130
140 - 150 - 160 - 170 - 180 - 190 - 200 - 210 - 220 - 230 - 240 - 250 - 260
270 - 280 - 290 - 300 - 310 - 320 - 330 - 340 - 350 - 360 - 370 - 380 - 390
400 - 410 - 420 - 430 - 440 - 450 - 460 - 470 - 480 - 490 - 500 - 510 - 520
530 - 540 - 550 - 560 - 570 - 580 - 590 - 600 - 610 - 620 - 630 - 640 - 650
660 - 670 - 680 - 690 - 700 - 710 - 720 - 730 - 740 - 750 - 760 - 770 - 780
790 - 800 - 810 - 820 - 830 - 840 - 850 - 860 - 870 - 880 - 890 - 900 - 910
920 - 930 - 940 - 950 - 960 - 970 - 980 - 990 - 1000

Notes: ______________________________

____

____

____

Reg Reps: 10 - 20 - 30 - 40 - 50 - 60 - 70 - 80 - 90 - 100 - 110 - 120 - 130
140 - 150 - 160 - 170 - 180 - 190 - 200 - 210 - 220 - 230 - 240 - 250 - 260
270 - 280 - 290 - 300 - 310 - 320 - 330 - 340 - 350 - 360 - 370 - 380 - 390
400 - 410 - 420 - 430 - 440 - 450 - 460 - 470 - 480 - 490 - 500 - 510 - 520
530 - 540 - 550 - 560 - 570 - 580 - 590 - 600 - 610 - 620 - 630 - 640 - 650
660 - 670 - 680 - 690 - 700 - 710 - 720 - 730 - 740 - 750 - 760 - 770 - 780
790 - 800 - 810 - 820 - 830 - 840 - 850 - 860 - 870 - 880 - 890 - 900 - 910
920 - 930 - 940 - 950 - 960 - 970 - 980 - 990 - 1000

Notes: ______________________________

____

____

______________________________ **End Date:** ____________

PICK HANDGUN UP FROM THE GROUND: NON-DOMINANT HAND ONLY

Holster Position:______________ **Hand:** ______ **Start Date:** _________
Spare Magazine Position: ____________ **Firearm:** _________________
Cover Garment: __________________________________ **Caliber:** ______
Notes:____

____

____

Slow Reps: 10 - 20 - 30 - 40 - 50 - 60 - 70 - 80 - 90 - 100 - 110 - 120 - 130
140 - 150 - 160 - 170 - 180 - 190 - 200 - 210 - 220 - 230 - 240 - 250 - 260
270 - 280 - 290 - 300 - 310 - 320 - 330 - 340 - 350 - 360 - 370 - 380 - 390
400 - 410 - 420 - 430 - 440 - 450 - 460 - 470 - 480 - 490 - 500 - 510 - 520
530 - 540 - 550 - 560 - 570 - 580 - 590 - 600 - 610 - 620 - 630 - 640 - 650
660 - 670 - 680 - 690 - 700 - 710 - 720 - 730 - 740 - 750 - 760 - 770 - 780
790 - 800 - 810 - 820 - 830 - 840 - 850 - 860 - 870 - 880 - 890 - 900 - 910
920 - 930 - 940 - 950 - 960 - 970 - 980 - 990 - 1000

Reg Reps: 10 - 20 - 30 - 40 - 50 - 60 - 70 - 80 - 90 - 100 - 110 - 120 - 130
140 - 150 - 160 - 170 - 180 - 190 - 200 - 210 - 220 - 230 - 240 - 250 - 260
270 - 280 - 290 - 300 - 310 - 320 - 330 - 340 - 350 - 360 - 370 - 380 - 390
400 - 410 - 420 - 430 - 440 - 450 - 460 - 470 - 480 - 490 - 500 - 510 - 520
530 - 540 - 550 - 560 - 570 - 580 - 590 - 600 - 610 - 620 - 630 - 640 - 650
660 - 670 - 680 - 690 - 700 - 710 - 720 - 730 - 740 - 750 - 760 - 770 - 780
790 - 800 - 810 - 820 - 830 - 840 - 850 - 860 - 870 - 880 - 890 - 900 - 910
920 - 930 - 940 - 950 - 960 - 970 - 980 - 990 - 1000

Notes: ____

____

____

Reg Reps: 10 - 20 - 30 - 40 - 50 - 60 - 70 - 80 - 90 - 100 - 110 - 120 - 130
140 - 150 - 160 - 170 - 180 - 190 - 200 - 210 - 220 - 230 - 240 - 250 - 260
270 - 280 - 290 - 300 - 310 - 320 - 330 - 340 - 350 - 360 - 370 - 380 - 390
400 - 410 - 420 - 430 - 440 - 450 - 460 - 470 - 480 - 490 - 500 - 510 - 520
530 - 540 - 550 - 560 - 570 - 580 - 590 - 600 - 610 - 620 - 630 - 640 - 650
660 - 670 - 680 - 690 - 700 - 710 - 720 - 730 - 740 - 750 - 760 - 770 - 780
790 - 800 - 810 - 820 - 830 - 840 - 850 - 860 - 870 - 880 - 890 - 900 - 910
920 - 930 - 940 - 950 - 960 - 970 - 980 - 990 - 1000

Notes: ____

____

____

Reg Reps: 10 - 20 - 30 - 40 - 50 - 60 - 70 - 80 - 90 - 100 - 110 - 120 - 130 - 140 - 150 - 160 - 170 - 180 - 190 - 200 - 210 - 220 - 230 - 240 - 250 - 260 - 270 - 280 - 290 - 300 - 310 - 320 - 330 - 340 - 350 - 360 - 370 - 380 - 390 - 400 - 410 - 420 - 430 - 440 - 450 - 460 - 470 - 480 - 490 - 500 - 510 - 520 - 530 - 540 - 550 - 560 - 570 - 580 - 590 - 600 - 610 - 620 - 630 - 640 - 650 - 660 - 670 - 680 - 690 - 700 - 710 - 720 - 730 - 740 - 750 - 760 - 770 - 780 - 790 - 800 - 810 - 820 - 830 - 840 - 850 - 860 - 870 - 880 - 890 - 900 - 910 - 920 - 930 - 940 - 950 - 960 - 970 - 980 - 990 - 1000

Notes: ______________________________________

____

____

____

Reg Reps: 10 - 20 - 30 - 40 - 50 - 60 - 70 - 80 - 90 - 100 - 110 - 120 - 130 - 140 - 150 - 160 - 170 - 180 - 190 - 200 - 210 - 220 - 230 - 240 - 250 - 260 - 270 - 280 - 290 - 300 - 310 - 320 - 330 - 340 - 350 - 360 - 370 - 380 - 390 - 400 - 410 - 420 - 430 - 440 - 450 - 460 - 470 - 480 - 490 - 500 - 510 - 520 - 530 - 540 - 550 - 560 - 570 - 580 - 590 - 600 - 610 - 620 - 630 - 640 - 650 - 660 - 670 - 680 - 690 - 700 - 710 - 720 - 730 - 740 - 750 - 760 - 770 - 780 - 790 - 800 - 810 - 820 - 830 - 840 - 850 - 860 - 870 - 880 - 890 - 900 - 910 - 920 - 930 - 940 - 950 - 960 - 970 - 980 - 990 - 1000

Notes: ______________________________________

____

____

____

Reg Reps: 10 - 20 - 30 - 40 - 50 - 60 - 70 - 80 - 90 - 100 - 110 - 120 - 130 - 140 - 150 - 160 - 170 - 180 - 190 - 200 - 210 - 220 - 230 - 240 - 250 - 260 - 270 - 280 - 290 - 300 - 310 - 320 - 330 - 340 - 350 - 360 - 370 - 380 - 390 - 400 - 410 - 420 - 430 - 440 - 450 - 460 - 470 - 480 - 490 - 500 - 510 - 520 - 530 - 540 - 550 - 560 - 570 - 580 - 590 - 600 - 610 - 620 - 630 - 640 - 650 - 660 - 670 - 680 - 690 - 700 - 710 - 720 - 730 - 740 - 750 - 760 - 770 - 780 - 790 - 800 - 810 - 820 - 830 - 840 - 850 - 860 - 870 - 880 - 890 - 900 - 910 - 920 - 930 - 940 - 950 - 960 - 970 - 980 - 990 - 1000

Notes: ______________________________________

____

____

______________________________ **End Date:** __________

MY DRY-FIRE DRILL #1:______________________________

Holster Position:____________ **Hand:** _____ **Start Date:** ________
Spare Magazine Position: __________ **Firearm:** _______________
Cover Garment: ____________________________ **Caliber:** _____
Notes:____

Slow Reps: 10 - 20 - 30 - 40 - 50 - 60 - 70 - 80 - 90 - 100 - 110 - 120 - 130
140 - 150 - 160 - 170 - 180 - 190 - 200 - 210 - 220 - 230 - 240 - 250 - 260
270 - 280 - 290 - 300 - 310 - 320 - 330 - 340 - 350 - 360 - 370 - 380 - 390
400 - 410 - 420 - 430 - 440 - 450 - 460 - 470 - 480 - 490 - 500 - 510 - 520
530 - 540 - 550 - 560 - 570 - 580 - 590 - 600 - 610 - 620 - 630 - 640 - 650
660 - 670 - 680 - 690 - 700 - 710 - 720 - 730 - 740 - 750 - 760 - 770 - 780
790 - 800 - 810 - 820 - 830 - 840 - 850 - 860 - 870 - 880 - 890 - 900 - 910
920 - 930 - 940 - 950 - 960 - 970 - 980 - 990 - 1000

Reg Reps: 10 - 20 - 30 - 40 - 50 - 60 - 70 - 80 - 90 - 100 - 110 - 120 - 130
140 - 150 - 160 - 170 - 180 - 190 - 200 - 210 - 220 - 230 - 240 - 250 - 260
270 - 280 - 290 - 300 - 310 - 320 - 330 - 340 - 350 - 360 - 370 - 380 - 390
400 - 410 - 420 - 430 - 440 - 450 - 460 - 470 - 480 - 490 - 500 - 510 - 520
530 - 540 - 550 - 560 - 570 - 580 - 590 - 600 - 610 - 620 - 630 - 640 - 650
660 - 670 - 680 - 690 - 700 - 710 - 720 - 730 - 740 - 750 - 760 - 770 - 780
790 - 800 - 810 - 820 - 830 - 840 - 850 - 860 - 870 - 880 - 890 - 900 - 910
920 - 930 - 940 - 950 - 960 - 970 - 980 - 990 - 1000
Notes: ____

Reg Reps: 10 - 20 - 30 - 40 - 50 - 60 - 70 - 80 - 90 - 100 - 110 - 120 - 130
140 - 150 - 160 - 170 - 180 - 190 - 200 - 210 - 220 - 230 - 240 - 250 - 260
270 - 280 - 290 - 300 - 310 - 320 - 330 - 340 - 350 - 360 - 370 - 380 - 390
400 - 410 - 420 - 430 - 440 - 450 - 460 - 470 - 480 - 490 - 500 - 510 - 520
530 - 540 - 550 - 560 - 570 - 580 - 590 - 600 - 610 - 620 - 630 - 640 - 650
660 - 670 - 680 - 690 - 700 - 710 - 720 - 730 - 740 - 750 - 760 - 770 - 780
790 - 800 - 810 - 820 - 830 - 840 - 850 - 860 - 870 - 880 - 890 - 900 - 910
920 - 930 - 940 - 950 - 960 - 970 - 980 - 990 - 1000
Notes: ____

Reg Reps: 10 - 20 - 30 - 40 - 50 - 60 - 70 - 80 - 90 - 100 - 110 - 120 - 130
140 - 150 - 160 - 170 - 180 - 190 - 200 - 210 - 220 - 230 - 240 - 250 - 260
270 - 280 - 290 - 300 - 310 - 320 - 330 - 340 - 350 - 360 - 370 - 380 - 390
400 - 410 - 420 - 430 - 440 - 450 - 460 - 470 - 480 - 490 - 500 - 510 - 520
530 - 540 - 550 - 560 - 570 - 580 - 590 - 600 - 610 - 620 - 630 - 640 - 650
660 - 670 - 680 - 690 - 700 - 710 - 720 - 730 - 740 - 750 - 760 - 770 - 780
790 - 800 - 810 - 820 - 830 - 840 - 850 - 860 - 870 - 880 - 890 - 900 - 910
920 - 930 - 940 - 950 - 960 - 970 - 980 - 990 - 1000

Notes: ____

Reg Reps: 10 - 20 - 30 - 40 - 50 - 60 - 70 - 80 - 90 - 100 - 110 - 120 - 130
140 - 150 - 160 - 170 - 180 - 190 - 200 - 210 - 220 - 230 - 240 - 250 - 260
270 - 280 - 290 - 300 - 310 - 320 - 330 - 340 - 350 - 360 - 370 - 380 - 390
400 - 410 - 420 - 430 - 440 - 450 - 460 - 470 - 480 - 490 - 500 - 510 - 520
530 - 540 - 550 - 560 - 570 - 580 - 590 - 600 - 610 - 620 - 630 - 640 - 650
660 - 670 - 680 - 690 - 700 - 710 - 720 - 730 - 740 - 750 - 760 - 770 - 780
790 - 800 - 810 - 820 - 830 - 840 - 850 - 860 - 870 - 880 - 890 - 900 - 910
920 - 930 - 940 - 950 - 960 - 970 - 980 - 990 - 1000

Notes: ____

Reg Reps: 10 - 20 - 30 - 40 - 50 - 60 - 70 - 80 - 90 - 100 - 110 - 120 - 130
140 - 150 - 160 - 170 - 180 - 190 - 200 - 210 - 220 - 230 - 240 - 250 - 260
270 - 280 - 290 - 300 - 310 - 320 - 330 - 340 - 350 - 360 - 370 - 380 - 390
400 - 410 - 420 - 430 - 440 - 450 - 460 - 470 - 480 - 490 - 500 - 510 - 520
530 - 540 - 550 - 560 - 570 - 580 - 590 - 600 - 610 - 620 - 630 - 640 - 650
660 - 670 - 680 - 690 - 700 - 710 - 720 - 730 - 740 - 750 - 760 - 770 - 780
790 - 800 - 810 - 820 - 830 - 840 - 850 - 860 - 870 - 880 - 890 - 900 - 910
920 - 930 - 940 - 950 - 960 - 970 - 980 - 990 - 1000

Notes: ____

End Date: ____________

MY DRY-FIRE DRILL #2: ___________________________________

Holster Position: _____________ **Hand:** ______ **Start Date:** ________
Spare Magazine Position: ___________ **Firearm:** _________________
Cover Garment: ________________________________ **Caliber:** ______
Notes: ____

____

____

Slow Reps: 10 - 20 - 30 - 40 - 50 - 60 - 70 - 80 - 90 - 100 - 110 - 120 - 130
140 - 150 - 160 - 170 - 180 - 190 - 200 - 210 - 220 - 230 - 240 - 250 - 260
270 - 280 - 290 - 300 - 310 - 320 - 330 - 340 - 350 - 360 - 370 - 380 - 390
400 - 410 - 420 - 430 - 440 - 450 - 460 - 470 - 480 - 490 - 500 - 510 - 520
530 - 540 - 550 - 560 - 570 - 580 - 590 - 600 - 610 - 620 - 630 - 640 - 650
660 - 670 - 680 - 690 - 700 - 710 - 720 - 730 - 740 - 750 - 760 - 770 - 780
790 - 800 - 810 - 820 - 830 - 840 - 850 - 860 - 870 - 880 - 890 - 900 - 910
920 - 930 - 940 - 950 - 960 - 970 - 980 - 990 - 1000

Reg Reps: 10 - 20 - 30 - 40 - 50 - 60 - 70 - 80 - 90 - 100 - 110 - 120 - 130
140 - 150 - 160 - 170 - 180 - 190 - 200 - 210 - 220 - 230 - 240 - 250 - 260
270 - 280 - 290 - 300 - 310 - 320 - 330 - 340 - 350 - 360 - 370 - 380 - 390
400 - 410 - 420 - 430 - 440 - 450 - 460 - 470 - 480 - 490 - 500 - 510 - 520
530 - 540 - 550 - 560 - 570 - 580 - 590 - 600 - 610 - 620 - 630 - 640 - 650
660 - 670 - 680 - 690 - 700 - 710 - 720 - 730 - 740 - 750 - 760 - 770 - 780
790 - 800 - 810 - 820 - 830 - 840 - 850 - 860 - 870 - 880 - 890 - 900 - 910
920 - 930 - 940 - 950 - 960 - 970 - 980 - 990 - 1000
Notes: ____

____

____

Reg Reps: 10 - 20 - 30 - 40 - 50 - 60 - 70 - 80 - 90 - 100 - 110 - 120 - 130
140 - 150 - 160 - 170 - 180 - 190 - 200 - 210 - 220 - 230 - 240 - 250 - 260
270 - 280 - 290 - 300 - 310 - 320 - 330 - 340 - 350 - 360 - 370 - 380 - 390
400 - 410 - 420 - 430 - 440 - 450 - 460 - 470 - 480 - 490 - 500 - 510 - 520
530 - 540 - 550 - 560 - 570 - 580 - 590 - 600 - 610 - 620 - 630 - 640 - 650
660 - 670 - 680 - 690 - 700 - 710 - 720 - 730 - 740 - 750 - 760 - 770 - 780
790 - 800 - 810 - 820 - 830 - 840 - 850 - 860 - 870 - 880 - 890 - 900 - 910
920 - 930 - 940 - 950 - 960 - 970 - 980 - 990 - 1000
Notes: ____

____

____

Reg Reps: 10 - 20 - 30 - 40 - 50 - 60 - 70 - 80 - 90 - 100 - 110 - 120 - 130
140 - 150 - 160 - 170 - 180 - 190 - 200 - 210 - 220 - 230 - 240 - 250 - 260
270 - 280 - 290 - 300 - 310 - 320 - 330 - 340 - 350 - 360 - 370 - 380 - 390
400 - 410 - 420 - 430 - 440 - 450 - 460 - 470 - 480 - 490 - 500 - 510 - 520
530 - 540 - 550 - 560 - 570 - 580 - 590 - 600 - 610 - 620 - 630 - 640 - 650
660 - 670 - 680 - 690 - 700 - 710 - 720 - 730 - 740 - 750 - 760 - 770 - 780
790 - 800 - 810 - 820 - 830 - 840 - 850 - 860 - 870 - 880 - 890 - 900 - 910
920 - 930 - 940 - 950 - 960 - 970 - 980 - 990 - 1000

Notes:

Reg Reps: 10 - 20 - 30 - 40 - 50 - 60 - 70 - 80 - 90 - 100 - 110 - 120 - 130
140 - 150 - 160 - 170 - 180 - 190 - 200 - 210 - 220 - 230 - 240 - 250 - 260
270 - 280 - 290 - 300 - 310 - 320 - 330 - 340 - 350 - 360 - 370 - 380 - 390
400 - 410 - 420 - 430 - 440 - 450 - 460 - 470 - 480 - 490 - 500 - 510 - 520
530 - 540 - 550 - 560 - 570 - 580 - 590 - 600 - 610 - 620 - 630 - 640 - 650
660 - 670 - 680 - 690 - 700 - 710 - 720 - 730 - 740 - 750 - 760 - 770 - 780
790 - 800 - 810 - 820 - 830 - 840 - 850 - 860 - 870 - 880 - 890 - 900 - 910
920 - 930 - 940 - 950 - 960 - 970 - 980 - 990 - 1000

Notes:

Reg Reps: 10 - 20 - 30 - 40 - 50 - 60 - 70 - 80 - 90 - 100 - 110 - 120 - 130
140 - 150 - 160 - 170 - 180 - 190 - 200 - 210 - 220 - 230 - 240 - 250 - 260
270 - 280 - 290 - 300 - 310 - 320 - 330 - 340 - 350 - 360 - 370 - 380 - 390
400 - 410 - 420 - 430 - 440 - 450 - 460 - 470 - 480 - 490 - 500 - 510 - 520
530 - 540 - 550 - 560 - 570 - 580 - 590 - 600 - 610 - 620 - 630 - 640 - 650
660 - 670 - 680 - 690 - 700 - 710 - 720 - 730 - 740 - 750 - 760 - 770 - 780
790 - 800 - 810 - 820 - 830 - 840 - 850 - 860 - 870 - 880 - 890 - 900 - 910
920 - 930 - 940 - 950 - 960 - 970 - 980 - 990 - 1000

Notes:

End Date:

MY DRY-FIRE DRILL #3: ______________________________

Holster Position: ____________ **Hand:** _____ **Start Date:** ________
Spare Magazine Position: __________ **Firearm:** ______________
Cover Garment: ______________________________ **Caliber:** _____
Notes: ____

____

____

Slow Reps: 10 - 20 - 30 - 40 - 50 - 60 - 70 - 80 - 90 - 100 - 110 - 120 - 130
140 - 150 - 160 - 170 - 180 - 190 - 200 - 210 - 220 - 230 - 240 - 250 - 260
270 - 280 - 290 - 300 - 310 - 320 - 330 - 340 - 350 - 360 - 370 - 380 - 390
400 - 410 - 420 - 430 - 440 - 450 - 460 - 470 - 480 - 490 - 500 - 510 - 520
530 - 540 - 550 - 560 - 570 - 580 - 590 - 600 - 610 - 620 - 630 - 640 - 650
660 - 670 - 680 - 690 - 700 - 710 - 720 - 730 - 740 - 750 - 760 - 770 - 780
790 - 800 - 810 - 820 - 830 - 840 - 850 - 860 - 870 - 880 - 890 - 900 - 910
920 - 930 - 940 - 950 - 960 - 970 - 980 - 990 - 1000

Reg Reps: 10 - 20 - 30 - 40 - 50 - 60 - 70 - 80 - 90 - 100 - 110 - 120 - 130
140 - 150 - 160 - 170 - 180 - 190 - 200 - 210 - 220 - 230 - 240 - 250 - 260
270 - 280 - 290 - 300 - 310 - 320 - 330 - 340 - 350 - 360 - 370 - 380 - 390
400 - 410 - 420 - 430 - 440 - 450 - 460 - 470 - 480 - 490 - 500 - 510 - 520
530 - 540 - 550 - 560 - 570 - 580 - 590 - 600 - 610 - 620 - 630 - 640 - 650
660 - 670 - 680 - 690 - 700 - 710 - 720 - 730 - 740 - 750 - 760 - 770 - 780
790 - 800 - 810 - 820 - 830 - 840 - 850 - 860 - 870 - 880 - 890 - 900 - 910
920 - 930 - 940 - 950 - 960 - 970 - 980 - 990 - 1000
Notes: ____

____

____

Reg Reps: 10 - 20 - 30 - 40 - 50 - 60 - 70 - 80 - 90 - 100 - 110 - 120 - 130
140 - 150 - 160 - 170 - 180 - 190 - 200 - 210 - 220 - 230 - 240 - 250 - 260
270 - 280 - 290 - 300 - 310 - 320 - 330 - 340 - 350 - 360 - 370 - 380 - 390
400 - 410 - 420 - 430 - 440 - 450 - 460 - 470 - 480 - 490 - 500 - 510 - 520
530 - 540 - 550 - 560 - 570 - 580 - 590 - 600 - 610 - 620 - 630 - 640 - 650
660 - 670 - 680 - 690 - 700 - 710 - 720 - 730 - 740 - 750 - 760 - 770 - 780
790 - 800 - 810 - 820 - 830 - 840 - 850 - 860 - 870 - 880 - 890 - 900 - 910
920 - 930 - 940 - 950 - 960 - 970 - 980 - 990 - 1000
Notes: ____

____

____

Reg Reps: 10 - 20 - 30 - 40 - 50 - 60 - 70 - 80 - 90 - 100 - 110 - 120 - 130
140 - 150 - 160 - 170 - 180 - 190 - 200 - 210 - 220 - 230 - 240 - 250 - 260
270 - 280 - 290 - 300 - 310 - 320 - 330 - 340 - 350 - 360 - 370 - 380 - 390
400 - 410 - 420 - 430 - 440 - 450 - 460 - 470 - 480 - 490 - 500 - 510 - 520
530 - 540 - 550 - 560 - 570 - 580 - 590 - 600 - 610 - 620 - 630 - 640 - 650
660 - 670 - 680 - 690 - 700 - 710 - 720 - 730 - 740 - 750 - 760 - 770 - 780
790 - 800 - 810 - 820 - 830 - 840 - 850 - 860 - 870 - 880 - 890 - 900 - 910
920 - 930 - 940 - 950 - 960 - 970 - 980 - 990 - 1000

Notes: ______________________________

____

____

____

Reg Reps: 10 - 20 - 30 - 40 - 50 - 60 - 70 - 80 - 90 - 100 - 110 - 120 - 130
140 - 150 - 160 - 170 - 180 - 190 - 200 - 210 - 220 - 230 - 240 - 250 - 260
270 - 280 - 290 - 300 - 310 - 320 - 330 - 340 - 350 - 360 - 370 - 380 - 390
400 - 410 - 420 - 430 - 440 - 450 - 460 - 470 - 480 - 490 - 500 - 510 - 520
530 - 540 - 550 - 560 - 570 - 580 - 590 - 600 - 610 - 620 - 630 - 640 - 650
660 - 670 - 680 - 690 - 700 - 710 - 720 - 730 - 740 - 750 - 760 - 770 - 780
790 - 800 - 810 - 820 - 830 - 840 - 850 - 860 - 870 - 880 - 890 - 900 - 910
920 - 930 - 940 - 950 - 960 - 970 - 980 - 990 - 1000

Notes: ______________________________

____

____

____

Reg Reps: 10 - 20 - 30 - 40 - 50 - 60 - 70 - 80 - 90 - 100 - 110 - 120 - 130
140 - 150 - 160 - 170 - 180 - 190 - 200 - 210 - 220 - 230 - 240 - 250 - 260
270 - 280 - 290 - 300 - 310 - 320 - 330 - 340 - 350 - 360 - 370 - 380 - 390
400 - 410 - 420 - 430 - 440 - 450 - 460 - 470 - 480 - 490 - 500 - 510 - 520
530 - 540 - 550 - 560 - 570 - 580 - 590 - 600 - 610 - 620 - 630 - 640 - 650
660 - 670 - 680 - 690 - 700 - 710 - 720 - 730 - 740 - 750 - 760 - 770 - 780
790 - 800 - 810 - 820 - 830 - 840 - 850 - 860 - 870 - 880 - 890 - 900 - 910
920 - 930 - 940 - 950 - 960 - 970 - 980 - 990 - 1000

Notes: ______________________________

____

____

______________________________ **End Date:** __________

MY DRY-FIRE DRILL #4: ______________________________

Holster Position: ____________ **Hand:** ______ **Start Date:** ________
Spare Magazine Position: ___________ **Firearm:** _________________
Cover Garment: ______________________________ **Caliber:** ______
Notes: ____

____

____

Slow Reps: 10 - 20 - 30 - 40 - 50 - 60 - 70 - 80 - 90 - 100 - 110 - 120 - 130
140 - 150 - 160 - 170 - 180 - 190 - 200 - 210 - 220 - 230 - 240 - 250 - 260
270 - 280 - 290 - 300 - 310 - 320 - 330 - 340 - 350 - 360 - 370 - 380 - 390
400 - 410 - 420 - 430 - 440 - 450 - 460 - 470 - 480 - 490 - 500 - 510 - 520
530 - 540 - 550 - 560 - 570 - 580 - 590 - 600 - 610 - 620 - 630 - 640 - 650
660 - 670 - 680 - 690 - 700 - 710 - 720 - 730 - 740 - 750 - 760 - 770 - 780
790 - 800 - 810 - 820 - 830 - 840 - 850 - 860 - 870 - 880 - 890 - 900 - 910
920 - 930 - 940 - 950 - 960 - 970 - 980 - 990 - 1000

Reg Reps: 10 - 20 - 30 - 40 - 50 - 60 - 70 - 80 - 90 - 100 - 110 - 120 - 130
140 - 150 - 160 - 170 - 180 - 190 - 200 - 210 - 220 - 230 - 240 - 250 - 260
270 - 280 - 290 - 300 - 310 - 320 - 330 - 340 - 350 - 360 - 370 - 380 - 390
400 - 410 - 420 - 430 - 440 - 450 - 460 - 470 - 480 - 490 - 500 - 510 - 520
530 - 540 - 550 - 560 - 570 - 580 - 590 - 600 - 610 - 620 - 630 - 640 - 650
660 - 670 - 680 - 690 - 700 - 710 - 720 - 730 - 740 - 750 - 760 - 770 - 780
790 - 800 - 810 - 820 - 830 - 840 - 850 - 860 - 870 - 880 - 890 - 900 - 910
920 - 930 - 940 - 950 - 960 - 970 - 980 - 990 - 1000
Notes: ____

____

____

Reg Reps: 10 - 20 - 30 - 40 - 50 - 60 - 70 - 80 - 90 - 100 - 110 - 120 - 130
140 - 150 - 160 - 170 - 180 - 190 - 200 - 210 - 220 - 230 - 240 - 250 - 260
270 - 280 - 290 - 300 - 310 - 320 - 330 - 340 - 350 - 360 - 370 - 380 - 390
400 - 410 - 420 - 430 - 440 - 450 - 460 - 470 - 480 - 490 - 500 - 510 - 520
530 - 540 - 550 - 560 - 570 - 580 - 590 - 600 - 610 - 620 - 630 - 640 - 650
660 - 670 - 680 - 690 - 700 - 710 - 720 - 730 - 740 - 750 - 760 - 770 - 780
790 - 800 - 810 - 820 - 830 - 840 - 850 - 860 - 870 - 880 - 890 - 900 - 910
920 - 930 - 940 - 950 - 960 - 970 - 980 - 990 - 1000
Notes: ____

____

____

Reg Reps: 10 - 20 - 30 - 40 - 50 - 60 - 70 - 80 - 90 - 100 - 110 - 120 - 130
140 - 150 - 160 - 170 - 180 - 190 - 200 - 210 - 220 - 230 - 240 - 250 - 260
270 - 280 - 290 - 300 - 310 - 320 - 330 - 340 - 350 - 360 - 370 - 380 - 390
400 - 410 - 420 - 430 - 440 - 450 - 460 - 470 - 480 - 490 - 500 - 510 - 520
530 - 540 - 550 - 560 - 570 - 580 - 590 - 600 - 610 - 620 - 630 - 640 - 650
660 - 670 - 680 - 690 - 700 - 710 - 720 - 730 - 740 - 750 - 760 - 770 - 780
790 - 800 - 810 - 820 - 830 - 840 - 850 - 860 - 870 - 880 - 890 - 900 - 910
920 - 930 - 940 - 950 - 960 - 970 - 980 - 990 - 1000

Notes: ____

____

____

____

Reg Reps: 10 - 20 - 30 - 40 - 50 - 60 - 70 - 80 - 90 - 100 - 110 - 120 - 130
140 - 150 - 160 - 170 - 180 - 190 - 200 - 210 - 220 - 230 - 240 - 250 - 260
270 - 280 - 290 - 300 - 310 - 320 - 330 - 340 - 350 - 360 - 370 - 380 - 390
400 - 410 - 420 - 430 - 440 - 450 - 460 - 470 - 480 - 490 - 500 - 510 - 520
530 - 540 - 550 - 560 - 570 - 580 - 590 - 600 - 610 - 620 - 630 - 640 - 650
660 - 670 - 680 - 690 - 700 - 710 - 720 - 730 - 740 - 750 - 760 - 770 - 780
790 - 800 - 810 - 820 - 830 - 840 - 850 - 860 - 870 - 880 - 890 - 900 - 910
920 - 930 - 940 - 950 - 960 - 970 - 980 - 990 - 1000

Notes: ____

____

____

____

Reg Reps: 10 - 20 - 30 - 40 - 50 - 60 - 70 - 80 - 90 - 100 - 110 - 120 - 130
140 - 150 - 160 - 170 - 180 - 190 - 200 - 210 - 220 - 230 - 240 - 250 - 260
270 - 280 - 290 - 300 - 310 - 320 - 330 - 340 - 350 - 360 - 370 - 380 - 390
400 - 410 - 420 - 430 - 440 - 450 - 460 - 470 - 480 - 490 - 500 - 510 - 520
530 - 540 - 550 - 560 - 570 - 580 - 590 - 600 - 610 - 620 - 630 - 640 - 650
660 - 670 - 680 - 690 - 700 - 710 - 720 - 730 - 740 - 750 - 760 - 770 - 780
790 - 800 - 810 - 820 - 830 - 840 - 850 - 860 - 870 - 880 - 890 - 900 - 910
920 - 930 - 940 - 950 - 960 - 970 - 980 - 990 - 1000

Notes: ____

____

____

______________________________________ **End Date:** ____________

MY DRY-FIRE DRILL #5: ______________________________

Holster Position: ____________ **Hand:** _____ **Start Date:** ________
Spare Magazine Position: __________ **Firearm:** _______________
Cover Garment: ______________________________ **Caliber:** _____
Notes: ____
____
____

Slow Reps: 10 - 20 - 30 - 40 - 50 - 60 - 70 - 80 - 90 - 100 - 110 - 120 - 130
140 - 150 - 160 - 170 - 180 - 190 - 200 - 210 - 220 - 230 - 240 - 250 - 260
270 - 280 - 290 - 300 - 310 - 320 - 330 - 340 - 350 - 360 - 370 - 380 - 390
400 - 410 - 420 - 430 - 440 - 450 - 460 - 470 - 480 - 490 - 500 - 510 - 520
530 - 540 - 550 - 560 - 570 - 580 - 590 - 600 - 610 - 620 - 630 - 640 - 650
660 - 670 - 680 - 690 - 700 - 710 - 720 - 730 - 740 - 750 - 760 - 770 - 780
790 - 800 - 810 - 820 - 830 - 840 - 850 - 860 - 870 - 880 - 890 - 900 - 910
920 - 930 - 940 - 950 - 960 - 970 - 980 - 990 - 1000

Reg Reps: 10 - 20 - 30 - 40 - 50 - 60 - 70 - 80 - 90 - 100 - 110 - 120 - 130
140 - 150 - 160 - 170 - 180 - 190 - 200 - 210 - 220 - 230 - 240 - 250 - 260
270 - 280 - 290 - 300 - 310 - 320 - 330 - 340 - 350 - 360 - 370 - 380 - 390
400 - 410 - 420 - 430 - 440 - 450 - 460 - 470 - 480 - 490 - 500 - 510 - 520
530 - 540 - 550 - 560 - 570 - 580 - 590 - 600 - 610 - 620 - 630 - 640 - 650
660 - 670 - 680 - 690 - 700 - 710 - 720 - 730 - 740 - 750 - 760 - 770 - 780
790 - 800 - 810 - 820 - 830 - 840 - 850 - 860 - 870 - 880 - 890 - 900 - 910
920 - 930 - 940 - 950 - 960 - 970 - 980 - 990 - 1000
Notes: ____
____
____

Reg Reps: 10 - 20 - 30 - 40 - 50 - 60 - 70 - 80 - 90 - 100 - 110 - 120 - 130
140 - 150 - 160 - 170 - 180 - 190 - 200 - 210 - 220 - 230 - 240 - 250 - 260
270 - 280 - 290 - 300 - 310 - 320 - 330 - 340 - 350 - 360 - 370 - 380 - 390
400 - 410 - 420 - 430 - 440 - 450 - 460 - 470 - 480 - 490 - 500 - 510 - 520
530 - 540 - 550 - 560 - 570 - 580 - 590 - 600 - 610 - 620 - 630 - 640 - 650
660 - 670 - 680 - 690 - 700 - 710 - 720 - 730 - 740 - 750 - 760 - 770 - 780
790 - 800 - 810 - 820 - 830 - 840 - 850 - 860 - 870 - 880 - 890 - 900 - 910
920 - 930 - 940 - 950 - 960 - 970 - 980 - 990 - 1000
Notes: ____
____
____

Reg Reps: 10 - 20 - 30 - 40 - 50 - 60 - 70 - 80 - 90 - 100 - 110 - 120 - 130
140 - 150 - 160 - 170 - 180 - 190 - 200 - 210 - 220 - 230 - 240 - 250 - 260
270 - 280 - 290 - 300 - 310 - 320 - 330 - 340 - 350 - 360 - 370 - 380 - 390
400 - 410 - 420 - 430 - 440 - 450 - 460 - 470 - 480 - 490 - 500 - 510 - 520
530 - 540 - 550 - 560 - 570 - 580 - 590 - 600 - 610 - 620 - 630 - 640 - 650
660 - 670 - 680 - 690 - 700 - 710 - 720 - 730 - 740 - 750 - 760 - 770 - 780
790 - 800 - 810 - 820 - 830 - 840 - 850 - 860 - 870 - 880 - 890 - 900 - 910
920 - 930 - 940 - 950 - 960 - 970 - 980 - 990 - 1000

Notes: ______________________________

Reg Reps: 10 - 20 - 30 - 40 - 50 - 60 - 70 - 80 - 90 - 100 - 110 - 120 - 130
140 - 150 - 160 - 170 - 180 - 190 - 200 - 210 - 220 - 230 - 240 - 250 - 260
270 - 280 - 290 - 300 - 310 - 320 - 330 - 340 - 350 - 360 - 370 - 380 - 390
400 - 410 - 420 - 430 - 440 - 450 - 460 - 470 - 480 - 490 - 500 - 510 - 520
530 - 540 - 550 - 560 - 570 - 580 - 590 - 600 - 610 - 620 - 630 - 640 - 650
660 - 670 - 680 - 690 - 700 - 710 - 720 - 730 - 740 - 750 - 760 - 770 - 780
790 - 800 - 810 - 820 - 830 - 840 - 850 - 860 - 870 - 880 - 890 - 900 - 910
920 - 930 - 940 - 950 - 960 - 970 - 980 - 990 - 1000

Notes: ______________________________

Reg Reps: 10 - 20 - 30 - 40 - 50 - 60 - 70 - 80 - 90 - 100 - 110 - 120 - 130
140 - 150 - 160 - 170 - 180 - 190 - 200 - 210 - 220 - 230 - 240 - 250 - 260
270 - 280 - 290 - 300 - 310 - 320 - 330 - 340 - 350 - 360 - 370 - 380 - 390
400 - 410 - 420 - 430 - 440 - 450 - 460 - 470 - 480 - 490 - 500 - 510 - 520
530 - 540 - 550 - 560 - 570 - 580 - 590 - 600 - 610 - 620 - 630 - 640 - 650
660 - 670 - 680 - 690 - 700 - 710 - 720 - 730 - 740 - 750 - 760 - 770 - 780
790 - 800 - 810 - 820 - 830 - 840 - 850 - 860 - 870 - 880 - 890 - 900 - 910
920 - 930 - 940 - 950 - 960 - 970 - 980 - 990 - 1000

Notes: ______________________________

End Date: ____________

MY DRY-FIRE DRILL #6:____________________________________

Holster Position:_______________ **Hand:** ______ **Start Date:** _________
Spare Magazine Position: ____________ **Firearm:** __________________
Cover Garment: ___________________________________ **Caliber:** ______
Notes:____

Slow Reps: 10 - 20 - 30 - 40 - 50 - 60 - 70 - 80 - 90 - 100 - 110 - 120 - 130
140 - 150 - 160 - 170 - 180 - 190 - 200 - 210 - 220 - 230 - 240 - 250 - 260
270 - 280 - 290 - 300 - 310 - 320 - 330 - 340 - 350 - 360 - 370 - 380 - 390
400 - 410 - 420 - 430 - 440 - 450 - 460 - 470 - 480 - 490 - 500 - 510 - 520
530 - 540 - 550 - 560 - 570 - 580 - 590 - 600 - 610 - 620 - 630 - 640 - 650
660 - 670 - 680 - 690 - 700 - 710 - 720 - 730 - 740 - 750 - 760 - 770 - 780
790 - 800 - 810 - 820 - 830 - 840 - 850 - 860 - 870 - 880 - 890 - 900 - 910
920 - 930 - 940 - 950 - 960 - 970 - 980 - 990 - 1000

Reg Reps: 10 - 20 - 30 - 40 - 50 - 60 - 70 - 80 - 90 - 100 - 110 - 120 - 130
140 - 150 - 160 - 170 - 180 - 190 - 200 - 210 - 220 - 230 - 240 - 250 - 260
270 - 280 - 290 - 300 - 310 - 320 - 330 - 340 - 350 - 360 - 370 - 380 - 390
400 - 410 - 420 - 430 - 440 - 450 - 460 - 470 - 480 - 490 - 500 - 510 - 520
530 - 540 - 550 - 560 - 570 - 580 - 590 - 600 - 610 - 620 - 630 - 640 - 650
660 - 670 - 680 - 690 - 700 - 710 - 720 - 730 - 740 - 750 - 760 - 770 - 780
790 - 800 - 810 - 820 - 830 - 840 - 850 - 860 - 870 - 880 - 890 - 900 - 910
920 - 930 - 940 - 950 - 960 - 970 - 980 - 990 - 1000
Notes: ____

Reg Reps: 10 - 20 - 30 - 40 - 50 - 60 - 70 - 80 - 90 - 100 - 110 - 120 - 130
140 - 150 - 160 - 170 - 180 - 190 - 200 - 210 - 220 - 230 - 240 - 250 - 260
270 - 280 - 290 - 300 - 310 - 320 - 330 - 340 - 350 - 360 - 370 - 380 - 390
400 - 410 - 420 - 430 - 440 - 450 - 460 - 470 - 480 - 490 - 500 - 510 - 520
530 - 540 - 550 - 560 - 570 - 580 - 590 - 600 - 610 - 620 - 630 - 640 - 650
660 - 670 - 680 - 690 - 700 - 710 - 720 - 730 - 740 - 750 - 760 - 770 - 780
790 - 800 - 810 - 820 - 830 - 840 - 850 - 860 - 870 - 880 - 890 - 900 - 910
920 - 930 - 940 - 950 - 960 - 970 - 980 - 990 - 1000
Notes: ____

Reg Reps: 10 - 20 - 30 - 40 - 50 - 60 - 70 - 80 - 90 - 100 - 110 - 120 - 130
140 - 150 - 160 - 170 - 180 - 190 - 200 - 210 - 220 - 230 - 240 - 250 - 260
270 - 280 - 290 - 300 - 310 - 320 - 330 - 340 - 350 - 360 - 370 - 380 - 390
400 - 410 - 420 - 430 - 440 - 450 - 460 - 470 - 480 - 490 - 500 - 510 - 520
530 - 540 - 550 - 560 - 570 - 580 - 590 - 600 - 610 - 620 - 630 - 640 - 650
660 - 670 - 680 - 690 - 700 - 710 - 720 - 730 - 740 - 750 - 760 - 770 - 780
790 - 800 - 810 - 820 - 830 - 840 - 850 - 860 - 870 - 880 - 890 - 900 - 910
920 - 930 - 940 - 950 - 960 - 970 - 980 - 990 - 1000

Notes: ______________________________

____

____

____

Reg Reps: 10 - 20 - 30 - 40 - 50 - 60 - 70 - 80 - 90 - 100 - 110 - 120 - 130
140 - 150 - 160 - 170 - 180 - 190 - 200 - 210 - 220 - 230 - 240 - 250 - 260
270 - 280 - 290 - 300 - 310 - 320 - 330 - 340 - 350 - 360 - 370 - 380 - 390
400 - 410 - 420 - 430 - 440 - 450 - 460 - 470 - 480 - 490 - 500 - 510 - 520
530 - 540 - 550 - 560 - 570 - 580 - 590 - 600 - 610 - 620 - 630 - 640 - 650
660 - 670 - 680 - 690 - 700 - 710 - 720 - 730 - 740 - 750 - 760 - 770 - 780
790 - 800 - 810 - 820 - 830 - 840 - 850 - 860 - 870 - 880 - 890 - 900 - 910
920 - 930 - 940 - 950 - 960 - 970 - 980 - 990 - 1000

Notes: ______________________________

____

____

____

Reg Reps: 10 - 20 - 30 - 40 - 50 - 60 - 70 - 80 - 90 - 100 - 110 - 120 - 130
140 - 150 - 160 - 170 - 180 - 190 - 200 - 210 - 220 - 230 - 240 - 250 - 260
270 - 280 - 290 - 300 - 310 - 320 - 330 - 340 - 350 - 360 - 370 - 380 - 390
400 - 410 - 420 - 430 - 440 - 450 - 460 - 470 - 480 - 490 - 500 - 510 - 520
530 - 540 - 550 - 560 - 570 - 580 - 590 - 600 - 610 - 620 - 630 - 640 - 650
660 - 670 - 680 - 690 - 700 - 710 - 720 - 730 - 740 - 750 - 760 - 770 - 780
790 - 800 - 810 - 820 - 830 - 840 - 850 - 860 - 870 - 880 - 890 - 900 - 910
920 - 930 - 940 - 950 - 960 - 970 - 980 - 990 - 1000

Notes: ______________________________

____

____

______________________________ **End Date:** __________

MY DRY-FIRE DRILL #7: ______________________________

Holster Position: ____________ **Hand:** ______ **Start Date:** _________
Spare Magazine Position: ___________ **Firearm:** ________________
Cover Garment: ______________________________ **Caliber:** ______
Notes: ____

____

____

Slow Reps: 10 - 20 - 30 - 40 - 50 - 60 - 70 - 80 - 90 - 100 - 110 - 120 - 130
140 - 150 - 160 - 170 - 180 - 190 - 200 - 210 - 220 - 230 - 240 - 250 - 260
270 - 280 - 290 - 300 - 310 - 320 - 330 - 340 - 350 - 360 - 370 - 380 - 390
400 - 410 - 420 - 430 - 440 - 450 - 460 - 470 - 480 - 490 - 500 - 510 - 520
530 - 540 - 550 - 560 - 570 - 580 - 590 - 600 - 610 - 620 - 630 - 640 - 650
660 - 670 - 680 - 690 - 700 - 710 - 720 - 730 - 740 - 750 - 760 - 770 - 780
790 - 800 - 810 - 820 - 830 - 840 - 850 - 860 - 870 - 880 - 890 - 900 - 910
920 - 930 - 940 - 950 - 960 - 970 - 980 - 990 - 1000

Reg Reps: 10 - 20 - 30 - 40 - 50 - 60 - 70 - 80 - 90 - 100 - 110 - 120 - 130
140 - 150 - 160 - 170 - 180 - 190 - 200 - 210 - 220 - 230 - 240 - 250 - 260
270 - 280 - 290 - 300 - 310 - 320 - 330 - 340 - 350 - 360 - 370 - 380 - 390
400 - 410 - 420 - 430 - 440 - 450 - 460 - 470 - 480 - 490 - 500 - 510 - 520
530 - 540 - 550 - 560 - 570 - 580 - 590 - 600 - 610 - 620 - 630 - 640 - 650
660 - 670 - 680 - 690 - 700 - 710 - 720 - 730 - 740 - 750 - 760 - 770 - 780
790 - 800 - 810 - 820 - 830 - 840 - 850 - 860 - 870 - 880 - 890 - 900 - 910
920 - 930 - 940 - 950 - 960 - 970 - 980 - 990 - 1000
Notes: ____

____

____

Reg Reps: 10 - 20 - 30 - 40 - 50 - 60 - 70 - 80 - 90 - 100 - 110 - 120 - 130
140 - 150 - 160 - 170 - 180 - 190 - 200 - 210 - 220 - 230 - 240 - 250 - 260
270 - 280 - 290 - 300 - 310 - 320 - 330 - 340 - 350 - 360 - 370 - 380 - 390
400 - 410 - 420 - 430 - 440 - 450 - 460 - 470 - 480 - 490 - 500 - 510 - 520
530 - 540 - 550 - 560 - 570 - 580 - 590 - 600 - 610 - 620 - 630 - 640 - 650
660 - 670 - 680 - 690 - 700 - 710 - 720 - 730 - 740 - 750 - 760 - 770 - 780
790 - 800 - 810 - 820 - 830 - 840 - 850 - 860 - 870 - 880 - 890 - 900 - 910
920 - 930 - 940 - 950 - 960 - 970 - 980 - 990 - 1000
Notes: ____

____

____

Reg Reps: 10 - 20 - 30 - 40 - 50 - 60 - 70 - 80 - 90 - 100 - 110 - 120 - 130
140 - 150 - 160 - 170 - 180 - 190 - 200 - 210 - 220 - 230 - 240 - 250 - 260
270 - 280 - 290 - 300 - 310 - 320 - 330 - 340 - 350 - 360 - 370 - 380 - 390
400 - 410 - 420 - 430 - 440 - 450 - 460 - 470 - 480 - 490 - 500 - 510 - 520
530 - 540 - 550 - 560 - 570 - 580 - 590 - 600 - 610 - 620 - 630 - 640 - 650
660 - 670 - 680 - 690 - 700 - 710 - 720 - 730 - 740 - 750 - 760 - 770 - 780
790 - 800 - 810 - 820 - 830 - 840 - 850 - 860 - 870 - 880 - 890 - 900 - 910
920 - 930 - 940 - 950 - 960 - 970 - 980 - 990 - 1000

Notes: ______________________________

____

____

____

Reg Reps: 10 - 20 - 30 - 40 - 50 - 60 - 70 - 80 - 90 - 100 - 110 - 120 - 130
140 - 150 - 160 - 170 - 180 - 190 - 200 - 210 - 220 - 230 - 240 - 250 - 260
270 - 280 - 290 - 300 - 310 - 320 - 330 - 340 - 350 - 360 - 370 - 380 - 390
400 - 410 - 420 - 430 - 440 - 450 - 460 - 470 - 480 - 490 - 500 - 510 - 520
530 - 540 - 550 - 560 - 570 - 580 - 590 - 600 - 610 - 620 - 630 - 640 - 650
660 - 670 - 680 - 690 - 700 - 710 - 720 - 730 - 740 - 750 - 760 - 770 - 780
790 - 800 - 810 - 820 - 830 - 840 - 850 - 860 - 870 - 880 - 890 - 900 - 910
920 - 930 - 940 - 950 - 960 - 970 - 980 - 990 - 1000

Notes: ______________________________

____

____

____

Reg Reps: 10 - 20 - 30 - 40 - 50 - 60 - 70 - 80 - 90 - 100 - 110 - 120 - 130
140 - 150 - 160 - 170 - 180 - 190 - 200 - 210 - 220 - 230 - 240 - 250 - 260
270 - 280 - 290 - 300 - 310 - 320 - 330 - 340 - 350 - 360 - 370 - 380 - 390
400 - 410 - 420 - 430 - 440 - 450 - 460 - 470 - 480 - 490 - 500 - 510 - 520
530 - 540 - 550 - 560 - 570 - 580 - 590 - 600 - 610 - 620 - 630 - 640 - 650
660 - 670 - 680 - 690 - 700 - 710 - 720 - 730 - 740 - 750 - 760 - 770 - 780
790 - 800 - 810 - 820 - 830 - 840 - 850 - 860 - 870 - 880 - 890 - 900 - 910
920 - 930 - 940 - 950 - 960 - 970 - 980 - 990 - 1000

Notes: ______________________________

____

____

______________________________ **End Date:** ____________

MY DRY-FIRE DRILL #8:____________________________________

Holster Position:_____________ **Hand:** ______ **Start Date:** _________
Spare Magazine Position: ___________ **Firearm:** _________________
Cover Garment: _______________________________ **Caliber:** ______
Notes:____
____
____

Slow Reps: 10 - 20 - 30 - 40 - 50 - 60 - 70 - 80 - 90 - 100 - 110 - 120 - 130
140 - 150 - 160 - 170 - 180 - 190 - 200 - 210 - 220 - 230 - 240 - 250 - 260
270 - 280 - 290 - 300 - 310 - 320 - 330 - 340 - 350 - 360 - 370 - 380 - 390
400 - 410 - 420 - 430 - 440 - 450 - 460 - 470 - 480 - 490 - 500 - 510 - 520
530 - 540 - 550 - 560 - 570 - 580 - 590 - 600 - 610 - 620 - 630 - 640 - 650
660 - 670 - 680 - 690 - 700 - 710 - 720 - 730 - 740 - 750 - 760 - 770 - 780
790 - 800 - 810 - 820 - 830 - 840 - 850 - 860 - 870 - 880 - 890 - 900 - 910
920 - 930 - 940 - 950 - 960 - 970 - 980 - 990 - 1000

Reg Reps: 10 - 20 - 30 - 40 - 50 - 60 - 70 - 80 - 90 - 100 - 110 - 120 - 130
140 - 150 - 160 - 170 - 180 - 190 - 200 - 210 - 220 - 230 - 240 - 250 - 260
270 - 280 - 290 - 300 - 310 - 320 - 330 - 340 - 350 - 360 - 370 - 380 - 390
400 - 410 - 420 - 430 - 440 - 450 - 460 - 470 - 480 - 490 - 500 - 510 - 520
530 - 540 - 550 - 560 - 570 - 580 - 590 - 600 - 610 - 620 - 630 - 640 - 650
660 - 670 - 680 - 690 - 700 - 710 - 720 - 730 - 740 - 750 - 760 - 770 - 780
790 - 800 - 810 - 820 - 830 - 840 - 850 - 860 - 870 - 880 - 890 - 900 - 910
920 - 930 - 940 - 950 - 960 - 970 - 980 - 990 - 1000
Notes: ____
____
____

Reg Reps: 10 - 20 - 30 - 40 - 50 - 60 - 70 - 80 - 90 - 100 - 110 - 120 - 130
140 - 150 - 160 - 170 - 180 - 190 - 200 - 210 - 220 - 230 - 240 - 250 - 260
270 - 280 - 290 - 300 - 310 - 320 - 330 - 340 - 350 - 360 - 370 - 380 - 390
400 - 410 - 420 - 430 - 440 - 450 - 460 - 470 - 480 - 490 - 500 - 510 - 520
530 - 540 - 550 - 560 - 570 - 580 - 590 - 600 - 610 - 620 - 630 - 640 - 650
660 - 670 - 680 - 690 - 700 - 710 - 720 - 730 - 740 - 750 - 760 - 770 - 780
790 - 800 - 810 - 820 - 830 - 840 - 850 - 860 - 870 - 880 - 890 - 900 - 910
920 - 930 - 940 - 950 - 960 - 970 - 980 - 990 - 1000
Notes: ____
____
____

Reg Reps: 10 - 20 - 30 - 40 - 50 - 60 - 70 - 80 - 90 - 100 - 110 - 120 - 130
140 - 150 - 160 - 170 - 180 - 190 - 200 - 210 - 220 - 230 - 240 - 250 - 260
270 - 280 - 290 - 300 - 310 - 320 - 330 - 340 - 350 - 360 - 370 - 380 - 390
400 - 410 - 420 - 430 - 440 - 450 - 460 - 470 - 480 - 490 - 500 - 510 - 520
530 - 540 - 550 - 560 - 570 - 580 - 590 - 600 - 610 - 620 - 630 - 640 - 650
660 - 670 - 680 - 690 - 700 - 710 - 720 - 730 - 740 - 750 - 760 - 770 - 780
790 - 800 - 810 - 820 - 830 - 840 - 850 - 860 - 870 - 880 - 890 - 900 - 910
920 - 930 - 940 - 950 - 960 - 970 - 980 - 990 - 1000

Notes: ______________________________

____

____

____

Reg Reps: 10 - 20 - 30 - 40 - 50 - 60 - 70 - 80 - 90 - 100 - 110 - 120 - 130
140 - 150 - 160 - 170 - 180 - 190 - 200 - 210 - 220 - 230 - 240 - 250 - 260
270 - 280 - 290 - 300 - 310 - 320 - 330 - 340 - 350 - 360 - 370 - 380 - 390
400 - 410 - 420 - 430 - 440 - 450 - 460 - 470 - 480 - 490 - 500 - 510 - 520
530 - 540 - 550 - 560 - 570 - 580 - 590 - 600 - 610 - 620 - 630 - 640 - 650
660 - 670 - 680 - 690 - 700 - 710 - 720 - 730 - 740 - 750 - 760 - 770 - 780
790 - 800 - 810 - 820 - 830 - 840 - 850 - 860 - 870 - 880 - 890 - 900 - 910
920 - 930 - 940 - 950 - 960 - 970 - 980 - 990 - 1000

Notes: ______________________________

____

____

____

Reg Reps: 10 - 20 - 30 - 40 - 50 - 60 - 70 - 80 - 90 - 100 - 110 - 120 - 130
140 - 150 - 160 - 170 - 180 - 190 - 200 - 210 - 220 - 230 - 240 - 250 - 260
270 - 280 - 290 - 300 - 310 - 320 - 330 - 340 - 350 - 360 - 370 - 380 - 390
400 - 410 - 420 - 430 - 440 - 450 - 460 - 470 - 480 - 490 - 500 - 510 - 520
530 - 540 - 550 - 560 - 570 - 580 - 590 - 600 - 610 - 620 - 630 - 640 - 650
660 - 670 - 680 - 690 - 700 - 710 - 720 - 730 - 740 - 750 - 760 - 770 - 780
790 - 800 - 810 - 820 - 830 - 840 - 850 - 860 - 870 - 880 - 890 - 900 - 910
920 - 930 - 940 - 950 - 960 - 970 - 980 - 990 - 1000

Notes: ______________________________

____

____

______________________________ **End Date:** ____________

MY DRY-FIRE DRILL #9:______________________________

Holster Position:____________ **Hand:** ______ **Start Date:** ________
Spare Magazine Position: __________ **Firearm:** ________________
Cover Garment: ____________________________ **Caliber:** ______
Notes:____
____
____

Slow Reps: 10 - 20 - 30 - 40 - 50 - 60 - 70 - 80 - 90 - 100 - 110 - 120 - 130
140 - 150 - 160 - 170 - 180 - 190 - 200 - 210 - 220 - 230 - 240 - 250 - 260
270 - 280 - 290 - 300 - 310 - 320 - 330 - 340 - 350 - 360 - 370 - 380 - 390
400 - 410 - 420 - 430 - 440 - 450 - 460 - 470 - 480 - 490 - 500 - 510 - 520
530 - 540 - 550 - 560 - 570 - 580 - 590 - 600 - 610 - 620 - 630 - 640 - 650
660 - 670 - 680 - 690 - 700 - 710 - 720 - 730 - 740 - 750 - 760 - 770 - 780
790 - 800 - 810 - 820 - 830 - 840 - 850 - 860 - 870 - 880 - 890 - 900 - 910
920 - 930 - 940 - 950 - 960 - 970 - 980 - 990 - 1000

Reg Reps: 10 - 20 - 30 - 40 - 50 - 60 - 70 - 80 - 90 - 100 - 110 - 120 - 130
140 - 150 - 160 - 170 - 180 - 190 - 200 - 210 - 220 - 230 - 240 - 250 - 260
270 - 280 - 290 - 300 - 310 - 320 - 330 - 340 - 350 - 360 - 370 - 380 - 390
400 - 410 - 420 - 430 - 440 - 450 - 460 - 470 - 480 - 490 - 500 - 510 - 520
530 - 540 - 550 - 560 - 570 - 580 - 590 - 600 - 610 - 620 - 630 - 640 - 650
660 - 670 - 680 - 690 - 700 - 710 - 720 - 730 - 740 - 750 - 760 - 770 - 780
790 - 800 - 810 - 820 - 830 - 840 - 850 - 860 - 870 - 880 - 890 - 900 - 910
920 - 930 - 940 - 950 - 960 - 970 - 980 - 990 - 1000
Notes: ____
____
____

Reg Reps: 10 - 20 - 30 - 40 - 50 - 60 - 70 - 80 - 90 - 100 - 110 - 120 - 130
140 - 150 - 160 - 170 - 180 - 190 - 200 - 210 - 220 - 230 - 240 - 250 - 260
270 - 280 - 290 - 300 - 310 - 320 - 330 - 340 - 350 - 360 - 370 - 380 - 390
400 - 410 - 420 - 430 - 440 - 450 - 460 - 470 - 480 - 490 - 500 - 510 - 520
530 - 540 - 550 - 560 - 570 - 580 - 590 - 600 - 610 - 620 - 630 - 640 - 650
660 - 670 - 680 - 690 - 700 - 710 - 720 - 730 - 740 - 750 - 760 - 770 - 780
790 - 800 - 810 - 820 - 830 - 840 - 850 - 860 - 870 - 880 - 890 - 900 - 910
920 - 930 - 940 - 950 - 960 - 970 - 980 - 990 - 1000
Notes: ____
____
____

Reg Reps: 10 - 20 - 30 - 40 - 50 - 60 - 70 - 80 - 90 - 100 - 110 - 120 - 130
140 - 150 - 160 - 170 - 180 - 190 - 200 - 210 - 220 - 230 - 240 - 250 - 260
270 - 280 - 290 - 300 - 310 - 320 - 330 - 340 - 350 - 360 - 370 - 380 - 390
400 - 410 - 420 - 430 - 440 - 450 - 460 - 470 - 480 - 490 - 500 - 510 - 520
530 - 540 - 550 - 560 - 570 - 580 - 590 - 600 - 610 - 620 - 630 - 640 - 650
660 - 670 - 680 - 690 - 700 - 710 - 720 - 730 - 740 - 750 - 760 - 770 - 780
790 - 800 - 810 - 820 - 830 - 840 - 850 - 860 - 870 - 880 - 890 - 900 - 910
920 - 930 - 940 - 950 - 960 - 970 - 980 - 990 - 1000

Notes: ______________________________

____

____

____

Reg Reps: 10 - 20 - 30 - 40 - 50 - 60 - 70 - 80 - 90 - 100 - 110 - 120 - 130
140 - 150 - 160 - 170 - 180 - 190 - 200 - 210 - 220 - 230 - 240 - 250 - 260
270 - 280 - 290 - 300 - 310 - 320 - 330 - 340 - 350 - 360 - 370 - 380 - 390
400 - 410 - 420 - 430 - 440 - 450 - 460 - 470 - 480 - 490 - 500 - 510 - 520
530 - 540 - 550 - 560 - 570 - 580 - 590 - 600 - 610 - 620 - 630 - 640 - 650
660 - 670 - 680 - 690 - 700 - 710 - 720 - 730 - 740 - 750 - 760 - 770 - 780
790 - 800 - 810 - 820 - 830 - 840 - 850 - 860 - 870 - 880 - 890 - 900 - 910
920 - 930 - 940 - 950 - 960 - 970 - 980 - 990 - 1000

Notes: ______________________________

____

____

____

Reg Reps: 10 - 20 - 30 - 40 - 50 - 60 - 70 - 80 - 90 - 100 - 110 - 120 - 130
140 - 150 - 160 - 170 - 180 - 190 - 200 - 210 - 220 - 230 - 240 - 250 - 260
270 - 280 - 290 - 300 - 310 - 320 - 330 - 340 - 350 - 360 - 370 - 380 - 390
400 - 410 - 420 - 430 - 440 - 450 - 460 - 470 - 480 - 490 - 500 - 510 - 520
530 - 540 - 550 - 560 - 570 - 580 - 590 - 600 - 610 - 620 - 630 - 640 - 650
660 - 670 - 680 - 690 - 700 - 710 - 720 - 730 - 740 - 750 - 760 - 770 - 780
790 - 800 - 810 - 820 - 830 - 840 - 850 - 860 - 870 - 880 - 890 - 900 - 910
920 - 930 - 940 - 950 - 960 - 970 - 980 - 990 - 1000

Notes: ______________________________

____

____

______________________________ **End Date:** __________

MY DRY-FIRE DRILL #10: ______________________________

Holster Position: ____________ **Hand:** ______ **Start Date:** ________
Spare Magazine Position: __________ **Firearm:** ________________
Cover Garment: _____________________________ **Caliber:** ______
Notes: ____

____

____

Slow Reps: 10 - 20 - 30 - 40 - 50 - 60 - 70 - 80 - 90 - 100 - 110 - 120 - 130
140 - 150 - 160 - 170 - 180 - 190 - 200 - 210 - 220 - 230 - 240 - 250 - 260
270 - 280 - 290 - 300 - 310 - 320 - 330 - 340 - 350 - 360 - 370 - 380 - 390
400 - 410 - 420 - 430 - 440 - 450 - 460 - 470 - 480 - 490 - 500 - 510 - 520
530 - 540 - 550 - 560 - 570 - 580 - 590 - 600 - 610 - 620 - 630 - 640 - 650
660 - 670 - 680 - 690 - 700 - 710 - 720 - 730 - 740 - 750 - 760 - 770 - 780
790 - 800 - 810 - 820 - 830 - 840 - 850 - 860 - 870 - 880 - 890 - 900 - 910
920 - 930 - 940 - 950 - 960 - 970 - 980 - 990 - 1000

Reg Reps: 10 - 20 - 30 - 40 - 50 - 60 - 70 - 80 - 90 - 100 - 110 - 120 - 130
140 - 150 - 160 - 170 - 180 - 190 - 200 - 210 - 220 - 230 - 240 - 250 - 260
270 - 280 - 290 - 300 - 310 - 320 - 330 - 340 - 350 - 360 - 370 - 380 - 390
400 - 410 - 420 - 430 - 440 - 450 - 460 - 470 - 480 - 490 - 500 - 510 - 520
530 - 540 - 550 - 560 - 570 - 580 - 590 - 600 - 610 - 620 - 630 - 640 - 650
660 - 670 - 680 - 690 - 700 - 710 - 720 - 730 - 740 - 750 - 760 - 770 - 780
790 - 800 - 810 - 820 - 830 - 840 - 850 - 860 - 870 - 880 - 890 - 900 - 910
920 - 930 - 940 - 950 - 960 - 970 - 980 - 990 - 1000
Notes: ____

____

____

Reg Reps: 10 - 20 - 30 - 40 - 50 - 60 - 70 - 80 - 90 - 100 - 110 - 120 - 130
140 - 150 - 160 - 170 - 180 - 190 - 200 - 210 - 220 - 230 - 240 - 250 - 260
270 - 280 - 290 - 300 - 310 - 320 - 330 - 340 - 350 - 360 - 370 - 380 - 390
400 - 410 - 420 - 430 - 440 - 450 - 460 - 470 - 480 - 490 - 500 - 510 - 520
530 - 540 - 550 - 560 - 570 - 580 - 590 - 600 - 610 - 620 - 630 - 640 - 650
660 - 670 - 680 - 690 - 700 - 710 - 720 - 730 - 740 - 750 - 760 - 770 - 780
790 - 800 - 810 - 820 - 830 - 840 - 850 - 860 - 870 - 880 - 890 - 900 - 910
920 - 930 - 940 - 950 - 960 - 970 - 980 - 990 - 1000
Notes: ____

____

____

Reg Reps: 10 - 20 - 30 - 40 - 50 - 60 - 70 - 80 - 90 - 100 - 110 - 120 - 130 140 - 150 - 160 - 170 - 180 - 190 - 200 - 210 - 220 - 230 - 240 - 250 - 260 270 - 280 - 290 - 300 - 310 - 320 - 330 - 340 - 350 - 360 - 370 - 380 - 390 400 - 410 - 420 - 430 - 440 - 450 - 460 - 470 - 480 - 490 - 500 - 510 - 520 530 - 540 - 550 - 560 - 570 - 580 - 590 - 600 - 610 - 620 - 630 - 640 - 650 660 - 670 - 680 - 690 - 700 - 710 - 720 - 730 - 740 - 750 - 760 - 770 - 780 790 - 800 - 810 - 820 - 830 - 840 - 850 - 860 - 870 - 880 - 890 - 900 - 910 920 - 930 - 940 - 950 - 960 - 970 - 980 - 990 - 1000

Notes: ______________________________

____

____

____

Reg Reps: 10 - 20 - 30 - 40 - 50 - 60 - 70 - 80 - 90 - 100 - 110 - 120 - 130 140 - 150 - 160 - 170 - 180 - 190 - 200 - 210 - 220 - 230 - 240 - 250 - 260 270 - 280 - 290 - 300 - 310 - 320 - 330 - 340 - 350 - 360 - 370 - 380 - 390 400 - 410 - 420 - 430 - 440 - 450 - 460 - 470 - 480 - 490 - 500 - 510 - 520 530 - 540 - 550 - 560 - 570 - 580 - 590 - 600 - 610 - 620 - 630 - 640 - 650 660 - 670 - 680 - 690 - 700 - 710 - 720 - 730 - 740 - 750 - 760 - 770 - 780 790 - 800 - 810 - 820 - 830 - 840 - 850 - 860 - 870 - 880 - 890 - 900 - 910 920 - 930 - 940 - 950 - 960 - 970 - 980 - 990 - 1000

Notes: ______________________________

____

____

____

Reg Reps: 10 - 20 - 30 - 40 - 50 - 60 - 70 - 80 - 90 - 100 - 110 - 120 - 130 140 - 150 - 160 - 170 - 180 - 190 - 200 - 210 - 220 - 230 - 240 - 250 - 260 270 - 280 - 290 - 300 - 310 - 320 - 330 - 340 - 350 - 360 - 370 - 380 - 390 400 - 410 - 420 - 430 - 440 - 450 - 460 - 470 - 480 - 490 - 500 - 510 - 520 530 - 540 - 550 - 560 - 570 - 580 - 590 - 600 - 610 - 620 - 630 - 640 - 650 660 - 670 - 680 - 690 - 700 - 710 - 720 - 730 - 740 - 750 - 760 - 770 - 780 790 - 800 - 810 - 820 - 830 - 840 - 850 - 860 - 870 - 880 - 890 - 900 - 910 920 - 930 - 940 - 950 - 960 - 970 - 980 - 990 - 1000

Notes: ______________________________

____

____

______________________________ **End Date:** ____________

MY DRY-FIRE DRILL #11: ________________________

Holster Position: ____________ **Hand:** _____ **Start Date:** _________
Spare Magazine Position: __________ **Firearm:** ______________
Cover Garment: ____________________________ **Caliber:** _____
Notes: ____

____

____

Slow Reps: 10 - 20 - 30 - 40 - 50 - 60 - 70 - 80 - 90 - 100 - 110 - 120 - 130
140 - 150 - 160 - 170 - 180 - 190 - 200 - 210 - 220 - 230 - 240 - 250 - 260
270 - 280 - 290 - 300 - 310 - 320 - 330 - 340 - 350 - 360 - 370 - 380 - 390
400 - 410 - 420 - 430 - 440 - 450 - 460 - 470 - 480 - 490 - 500 - 510 - 520
530 - 540 - 550 - 560 - 570 - 580 - 590 - 600 - 610 - 620 - 630 - 640 - 650
660 - 670 - 680 - 690 - 700 - 710 - 720 - 730 - 740 - 750 - 760 - 770 - 780
790 - 800 - 810 - 820 - 830 - 840 - 850 - 860 - 870 - 880 - 890 - 900 - 910
920 - 930 - 940 - 950 - 960 - 970 - 980 - 990 - 1000

Reg Reps: 10 - 20 - 30 - 40 - 50 - 60 - 70 - 80 - 90 - 100 - 110 - 120 - 130
140 - 150 - 160 - 170 - 180 - 190 - 200 - 210 - 220 - 230 - 240 - 250 - 260
270 - 280 - 290 - 300 - 310 - 320 - 330 - 340 - 350 - 360 - 370 - 380 - 390
400 - 410 - 420 - 430 - 440 - 450 - 460 - 470 - 480 - 490 - 500 - 510 - 520
530 - 540 - 550 - 560 - 570 - 580 - 590 - 600 - 610 - 620 - 630 - 640 - 650
660 - 670 - 680 - 690 - 700 - 710 - 720 - 730 - 740 - 750 - 760 - 770 - 780
790 - 800 - 810 - 820 - 830 - 840 - 850 - 860 - 870 - 880 - 890 - 900 - 910
920 - 930 - 940 - 950 - 960 - 970 - 980 - 990 - 1000
Notes: ____

____

____

Reg Reps: 10 - 20 - 30 - 40 - 50 - 60 - 70 - 80 - 90 - 100 - 110 - 120 - 130
140 - 150 - 160 - 170 - 180 - 190 - 200 - 210 - 220 - 230 - 240 - 250 - 260
270 - 280 - 290 - 300 - 310 - 320 - 330 - 340 - 350 - 360 - 370 - 380 - 390
400 - 410 - 420 - 430 - 440 - 450 - 460 - 470 - 480 - 490 - 500 - 510 - 520
530 - 540 - 550 - 560 - 570 - 580 - 590 - 600 - 610 - 620 - 630 - 640 - 650
660 - 670 - 680 - 690 - 700 - 710 - 720 - 730 - 740 - 750 - 760 - 770 - 780
790 - 800 - 810 - 820 - 830 - 840 - 850 - 860 - 870 - 880 - 890 - 900 - 910
920 - 930 - 940 - 950 - 960 - 970 - 980 - 990 - 1000
Notes: ____

____

____

Reg Reps: 10 - 20 - 30 - 40 - 50 - 60 - 70 - 80 - 90 - 100 - 110 - 120 - 130
140 - 150 - 160 - 170 - 180 - 190 - 200 - 210 - 220 - 230 - 240 - 250 - 260
270 - 280 - 290 - 300 - 310 - 320 - 330 - 340 - 350 - 360 - 370 - 380 - 390
400 - 410 - 420 - 430 - 440 - 450 - 460 - 470 - 480 - 490 - 500 - 510 - 520
530 - 540 - 550 - 560 - 570 - 580 - 590 - 600 - 610 - 620 - 630 - 640 - 650
660 - 670 - 680 - 690 - 700 - 710 - 720 - 730 - 740 - 750 - 760 - 770 - 780
790 - 800 - 810 - 820 - 830 - 840 - 850 - 860 - 870 - 880 - 890 - 900 - 910
920 - 930 - 940 - 950 - 960 - 970 - 980 - 990 - 1000

Notes: ______________________________

Reg Reps: 10 - 20 - 30 - 40 - 50 - 60 - 70 - 80 - 90 - 100 - 110 - 120 - 130
140 - 150 - 160 - 170 - 180 - 190 - 200 - 210 - 220 - 230 - 240 - 250 - 260
270 - 280 - 290 - 300 - 310 - 320 - 330 - 340 - 350 - 360 - 370 - 380 - 390
400 - 410 - 420 - 430 - 440 - 450 - 460 - 470 - 480 - 490 - 500 - 510 - 520
530 - 540 - 550 - 560 - 570 - 580 - 590 - 600 - 610 - 620 - 630 - 640 - 650
660 - 670 - 680 - 690 - 700 - 710 - 720 - 730 - 740 - 750 - 760 - 770 - 780
790 - 800 - 810 - 820 - 830 - 840 - 850 - 860 - 870 - 880 - 890 - 900 - 910
920 - 930 - 940 - 950 - 960 - 970 - 980 - 990 - 1000

Notes: ______________________________

Reg Reps: 10 - 20 - 30 - 40 - 50 - 60 - 70 - 80 - 90 - 100 - 110 - 120 - 130
140 - 150 - 160 - 170 - 180 - 190 - 200 - 210 - 220 - 230 - 240 - 250 - 260
270 - 280 - 290 - 300 - 310 - 320 - 330 - 340 - 350 - 360 - 370 - 380 - 390
400 - 410 - 420 - 430 - 440 - 450 - 460 - 470 - 480 - 490 - 500 - 510 - 520
530 - 540 - 550 - 560 - 570 - 580 - 590 - 600 - 610 - 620 - 630 - 640 - 650
660 - 670 - 680 - 690 - 700 - 710 - 720 - 730 - 740 - 750 - 760 - 770 - 780
790 - 800 - 810 - 820 - 830 - 840 - 850 - 860 - 870 - 880 - 890 - 900 - 910
920 - 930 - 940 - 950 - 960 - 970 - 980 - 990 - 1000

Notes: ______________________________

End Date: ______________

MY DRY-FIRE DRILL #12:______________________________

Holster Position:____________ **Hand:** ______ **Start Date:** _________
Spare Magazine Position: ___________ **Firearm:** _______________
Cover Garment: ____________________________ **Caliber:** ______
Notes:____

____

____

Slow Reps: 10 - 20 - 30 - 40 - 50 - 60 - 70 - 80 - 90 - 100 - 110 - 120 - 130
140 - 150 - 160 - 170 - 180 - 190 - 200 - 210 - 220 - 230 - 240 - 250 - 260
270 - 280 - 290 - 300 - 310 - 320 - 330 - 340 - 350 - 360 - 370 - 380 - 390
400 - 410 - 420 - 430 - 440 - 450 - 460 - 470 - 480 - 490 - 500 - 510 - 520
530 - 540 - 550 - 560 - 570 - 580 - 590 - 600 - 610 - 620 - 630 - 640 - 650
660 - 670 - 680 - 690 - 700 - 710 - 720 - 730 - 740 - 750 - 760 - 770 - 780
790 - 800 - 810 - 820 - 830 - 840 - 850 - 860 - 870 - 880 - 890 - 900 - 910
920 - 930 - 940 - 950 - 960 - 970 - 980 - 990 - 1000

Reg Reps: 10 - 20 - 30 - 40 - 50 - 60 - 70 - 80 - 90 - 100 - 110 - 120 - 130
140 - 150 - 160 - 170 - 180 - 190 - 200 - 210 - 220 - 230 - 240 - 250 - 260
270 - 280 - 290 - 300 - 310 - 320 - 330 - 340 - 350 - 360 - 370 - 380 - 390
400 - 410 - 420 - 430 - 440 - 450 - 460 - 470 - 480 - 490 - 500 - 510 - 520
530 - 540 - 550 - 560 - 570 - 580 - 590 - 600 - 610 - 620 - 630 - 640 - 650
660 - 670 - 680 - 690 - 700 - 710 - 720 - 730 - 740 - 750 - 760 - 770 - 780
790 - 800 - 810 - 820 - 830 - 840 - 850 - 860 - 870 - 880 - 890 - 900 - 910
920 - 930 - 940 - 950 - 960 - 970 - 980 - 990 - 1000
Notes: ____

____

____

Reg Reps: 10 - 20 - 30 - 40 - 50 - 60 - 70 - 80 - 90 - 100 - 110 - 120 - 130
140 - 150 - 160 - 170 - 180 - 190 - 200 - 210 - 220 - 230 - 240 - 250 - 260
270 - 280 - 290 - 300 - 310 - 320 - 330 - 340 - 350 - 360 - 370 - 380 - 390
400 - 410 - 420 - 430 - 440 - 450 - 460 - 470 - 480 - 490 - 500 - 510 - 520
530 - 540 - 550 - 560 - 570 - 580 - 590 - 600 - 610 - 620 - 630 - 640 - 650
660 - 670 - 680 - 690 - 700 - 710 - 720 - 730 - 740 - 750 - 760 - 770 - 780
790 - 800 - 810 - 820 - 830 - 840 - 850 - 860 - 870 - 880 - 890 - 900 - 910
920 - 930 - 940 - 950 - 960 - 970 - 980 - 990 - 1000
Notes: ____

____

____

Reg Reps: 10 - 20 - 30 - 40 - 50 - 60 - 70 - 80 - 90 - 100 - 110 - 120 - 130
140 - 150 - 160 - 170 - 180 - 190 - 200 - 210 - 220 - 230 - 240 - 250 - 260
270 - 280 - 290 - 300 - 310 - 320 - 330 - 340 - 350 - 360 - 370 - 380 - 390
400 - 410 - 420 - 430 - 440 - 450 - 460 - 470 - 480 - 490 - 500 - 510 - 520
530 - 540 - 550 - 560 - 570 - 580 - 590 - 600 - 610 - 620 - 630 - 640 - 650
660 - 670 - 680 - 690 - 700 - 710 - 720 - 730 - 740 - 750 - 760 - 770 - 780
790 - 800 - 810 - 820 - 830 - 840 - 850 - 860 - 870 - 880 - 890 - 900 - 910
920 - 930 - 940 - 950 - 960 - 970 - 980 - 990 - 1000

Notes: ______________________________

Reg Reps: 10 - 20 - 30 - 40 - 50 - 60 - 70 - 80 - 90 - 100 - 110 - 120 - 130
140 - 150 - 160 - 170 - 180 - 190 - 200 - 210 - 220 - 230 - 240 - 250 - 260
270 - 280 - 290 - 300 - 310 - 320 - 330 - 340 - 350 - 360 - 370 - 380 - 390
400 - 410 - 420 - 430 - 440 - 450 - 460 - 470 - 480 - 490 - 500 - 510 - 520
530 - 540 - 550 - 560 - 570 - 580 - 590 - 600 - 610 - 620 - 630 - 640 - 650
660 - 670 - 680 - 690 - 700 - 710 - 720 - 730 - 740 - 750 - 760 - 770 - 780
790 - 800 - 810 - 820 - 830 - 840 - 850 - 860 - 870 - 880 - 890 - 900 - 910
920 - 930 - 940 - 950 - 960 - 970 - 980 - 990 - 1000

Notes: ______________________________

Reg Reps: 10 - 20 - 30 - 40 - 50 - 60 - 70 - 80 - 90 - 100 - 110 - 120 - 130
140 - 150 - 160 - 170 - 180 - 190 - 200 - 210 - 220 - 230 - 240 - 250 - 260
270 - 280 - 290 - 300 - 310 - 320 - 330 - 340 - 350 - 360 - 370 - 380 - 390
400 - 410 - 420 - 430 - 440 - 450 - 460 - 470 - 480 - 490 - 500 - 510 - 520
530 - 540 - 550 - 560 - 570 - 580 - 590 - 600 - 610 - 620 - 630 - 640 - 650
660 - 670 - 680 - 690 - 700 - 710 - 720 - 730 - 740 - 750 - 760 - 770 - 780
790 - 800 - 810 - 820 - 830 - 840 - 850 - 860 - 870 - 880 - 890 - 900 - 910
920 - 930 - 940 - 950 - 960 - 970 - 980 - 990 - 1000

Notes: ______________________________

End Date: __________

ADDITIONAL NOTES

Previous Page:_______ Drill:_______________________________ Rep #:_____

Previous Page:_______ Drill:_______________________________ Rep #:_____

Previous Page:_______ Drill:_______________________________ Rep #:_____

Previous Page:_______ Drill:_______________________________ Rep #:_____

ADDITIONAL NOTES

Previous Page:_______ Drill:________________________________ Rep #:_____

Previous Page:_______ Drill:________________________________ Rep #:_____

Previous Page:_______ Drill:________________________________ Rep #:_____

Previous Page:_______ Drill:________________________________ Rep #:_____

ADDITIONAL NOTES

Previous Page:_______ Drill:_________________________________ Rep #:_____

Previous Page:_______ Drill:_________________________________ Rep #:_____

Previous Page:_______ Drill:_________________________________ Rep #:_____

Previous Page:_______ Drill:_________________________________ Rep #:_______

ADDITIONAL NOTES

Previous Page:_______ **Drill:**________________________________ **Rep #:**______

Previous Page:_______ **Drill:**________________________________ **Rep #:**______

Previous Page:_______ **Drill:**________________________________ **Rep #:**______

Previous Page:_______ **Drill:**________________________________ **Rep #:**______

ADDITIONAL NOTES

Previous Page:_______ **Drill:**________________________________ **Rep #:**_____

Previous Page:_______ **Drill:**________________________________ **Rep #:**_____

Previous Page:_______ **Drill:**________________________________ **Rep #:**_____

Previous Page:_______ **Drill:**________________________________ **Rep #:**_____

ADDITIONAL NOTES

Previous Page:_______ Drill:________________________________ Rep #:_____

Previous Page:_______ Drill:________________________________ Rep #:_____

Previous Page:_______ Drill:________________________________ Rep #:_____

Previous Page:_______ Drill:________________________________ Rep #:_____

ADDITIONAL NOTES

Previous Page:_______ **Drill:**________________________________ **Rep #:**_____

Previous Page:_______ **Drill:**________________________________ **Rep #:**_____

Previous Page:_______ **Drill:**________________________________ **Rep #:**_____

Previous Page:_______ **Drill:**________________________________ **Rep #:**_____

ADDITIONAL NOTES

Previous Page:_______ Drill:________________________________ Rep #:_____

Previous Page:_______ Drill:________________________________ Rep #:_____

Previous Page:_______ Drill:________________________________ Rep #:_____

Previous Page:_______ Drill:________________________________ Rep #:_____

ADDITIONAL NOTES

Previous Page:_______ Drill:_________________________________ Rep #:_____

Previous Page:_______ Drill:_________________________________ Rep #:_____

Previous Page:_______ Drill:_________________________________ Rep #:_____

Previous Page:_______ Drill:_________________________________ Rep #:_____

ADDITIONAL NOTES

Previous Page:_______ Drill:________________________________ Rep #:______

Previous Page:_______ Drill:________________________________ Rep #:______

Previous Page:_______ Drill:________________________________ Rep #:______

Previous Page:_______ Drill:________________________________ Rep #:______

ADDITIONAL NOTES

Previous Page:_______ Drill:________________________________ Rep #:_____

Previous Page:_______ Drill:________________________________ Rep #:_____

Previous Page:_______ Drill:________________________________ Rep #:_____

Previous Page:_______ Drill:________________________________ Rep #:_____

ADDITIONAL NOTES

Previous Page:_______ Drill:____________________________________ Rep #:_____

Previous Page:_______ Drill:____________________________________ Rep #:_____

Previous Page:_______ Drill:____________________________________ Rep #:_____

Previous Page:_______ Drill:____________________________________ Rep #:_____

ADDITIONAL NOTES

Previous Page:_______ Drill:_________________________________ Rep #:_____

Previous Page:_______ Drill:_________________________________ Rep #:_____

Previous Page:_______ Drill:_________________________________ Rep #:_____

Previous Page:_______ Drill:_________________________________ Rep #:_____

ADDITIONAL NOTES

Previous Page:_______ **Drill:**________________________________ **Rep #:**_____

Previous Page:_______ **Drill:**________________________________ **Rep #:**_____

Previous Page:_______ **Drill:**________________________________ **Rep #:**_____

Previous Page:_______ **Drill:**________________________________ **Rep #:**_______

ADDITIONAL NOTES

Previous Page:_______ Drill:________________________________ Rep #:_____

Previous Page:_______ Drill:________________________________ Rep #:_____

Previous Page:_______ Drill:________________________________ Rep #:_____

Previous Page:_______ Drill:________________________________ Rep #:_______

ADDITIONAL NOTES

Previous Page:_______ Drill:_________________________________ Rep #:_____

Previous Page:_______ Drill:_________________________________ Rep #:_____

Previous Page:_______ Drill:_________________________________ Rep #:_____

Previous Page:_______ Drill:_________________________________ Rep #:_____

Made in the USA
Monee, IL
04 November 2026

67377449R00075